The Glaciers of the Alps & Mountaineering in 1861

THE GLACIERS OF THE ALPS & MOUNTAINEERING IN 1861

JOHN TYNALL

The Glaciers of the Alps & Mountaineering in 1861
© 2005 Cosimo, Inc.
All rights reserved. No part of this book may be used or reproduced in any manner whatsoever without prior written permission except in the case of brief quotations embodied in critical articles or reviews.
For information, address:

Cosimo, P.O. Box 416
Old Chelsea Station
New York, NY 10113-0416

or visit our website at:
www.cosimobooks.com

The Glaciers of the Alps & Mountaineering in 1861 originally published by E.P. Dutton & Company in 1906.

Library of Congress Cataloging-in-Publication Data
A catalog record for this book is available from the Library of Congress

Cover design by www.wiselephant.com

ISBN: 1-59605-175-2

CONTENTS

GLACIERS OF THE ALPS—PART I

	PAGE
INTRODUCTORY	3

EXPEDITION OF 1856—
    THE OBERLAND 10
    THE TYROL 22

EXPEDITION OF 1857—
    THE LAKE OF GENEVA 31
    CHAMOUNI AND THE MONTANVERT . . . 34
    THE MER DE GLACE 38
    THE JARDIN 54
    FIRST ASCENT OF MONT BLANC, 1857 . . 60

EXPEDITION OF 1858—
    PASSAGE OF THE STRAHLECK 83
    ASCENT OF THE FINSTERAARHORN, 1858 . . 92
    FIRST ASCENT OF MONTE ROSA, 1858 . . 108
    THE GÖRNER GRAT AND THE RIFFELHORN.
        MAGNETIC PHENOMENA 121
    SECOND ASCENT OF MONTE ROSA, 1858 . . 133
    SECOND ASCENT OF MONT BLANC, 1858 . . 156

WINTER EXPEDITION TO THE MER DE
    GLACE, 1859 172

MOUNTAINEERING IN 1861

CHAP.		PAGE
I.	London to Meyringen	197
II.	Meyringen to the Grimsel, by the Urbachthal and Gauli Glacier	203
III.	The Grimsel and the Æggischhorn	210
IV.	The Bel Alp	216
V.	Reflections	221
VI.	Ascent of the Weisshorn	227
VII.	The Descent	241
VIII.	The Motion of Glaciers	247
IX.	Sunrise on the Pines	253
X.	Inspection of the Matterhorn	255
XI.	Over the Moro	263
XII.	The Old Weissthor	267

EDITOR'S NOTE

In order to include Tyndall's delightful Alpine book of 1861, which forms a natural sequel to the first descriptive part of the "Glaciers" volume, it has been found necessary to omit the second part of that work. It is intended at a later stage to publish some of Tyndall's purely scientific essays and treatises in a separate volume in the series.

INTRODUCTION

IT is not in my power to do justice to Tyndall's scientific discoveries, which lie for the most part outside the range of my studies ; but having been honoured with his friendship for many years, it is a pleasure to me,—and alas, few now living are able to do so,—to speak of him as a man, of the stimulating and noble influence which he exercised over all who had the privilege of his friendship.

More than forty years ago—in the summer of 1861—I had the pleasure of spending a short holiday in Switzerland with Huxley and Tyndall. Tyndall and I ascended the Galenstock, and started with Benen (who afterwards lost his life on the Haut de Cry) up the Jungfrau, but were stopped by an accident to one of our porters, who fell into a deep crevasse, from which we had some difficulty in extricating him, as Tyndall has graphically described in his "Hours of Exercise in the Alps."

Again in 1865 I spent a fortnight at Zermatt with him and Hirst just after the terrible catastrophe on the Matterhorn, searching, or rather intending to search, for the body of poor Lord Francis Douglas, which, however, the weather rendered impossible. On another occasion we went together to Naples to study an eruption of Vesuvius. Nearer home I had the pleasure of many short expeditions with him ; we attended several British Associations together, and belonged to the X Club, which dined together once a month, and which comprised also Huxley, Hooker, Herbert Spencer, Spottiswoode, G. Busk, Hirst, and Frankland.

These men were all pre-eminent in their own departments. Hooker, the only one, alas, still alive, is *Botanicorum facile princeps*, and was President of the Royal Society, as also were

Huxley and Spottiswoode; five were Presidents of the British Association: Spencer held a high place as a philosopher.

The science of several was so abstruse, and they themselves were so devoted to it and so retiring, that great as their reputation was among scientific men, they were scarcely appreciated at their full value by the outside world. Hirst, for instance, among mathematicians; and Busk—than whom no man stood higher or was more loved among naturalists.

As exponents of science Huxley and Tyndall were unrivalled. When they lectured at the Royal Institution the theatre was unable to hold all who wished to come. In style they were very different. Huxley convinced his audience and compelled their assent: Tyndall carried them with him.

They could not help agreeing with Huxley, even if they did not wish to do so: they wished to agree with Tyndall, if they could.

Tyndall was indeed most lovable and gentle, though he could be roused by anything which seemed to him an injustice: he did not form an opinion lightly, but when he saw his way clearly, as for instance on Home Rule, he was as firm as a rock. A Liberal in politics, he became a staunch Liberal Unionist.

Tyndall, like most of his friends, was a reverent agnostic. He did not believe that the ultimate truths of the Universe could be expressed in words, or that our limited and finite intelligence could as yet comprehend them.

His Belfast Address, which would probably shock few thinkers now, gave great offence at the time. It was soon after the foundation of School Boards, when the "religious difficulty" was much to the front, and in the autumn of the year in which, as I have already mentioned, he and I went to Naples together. One of the Belfast papers, discussing the Address, contrasted Tyndall's outspoken utterances with Huxley's (supposed) reticence, and ended a somewhat uncharitable criticism by observing that they could give no better illustration of Huxley's wisdom and Tyndall's recklessness than by mentioning the simple fact that while Prof. Tyndall most incautiously entered the crater of Vesuvius during an eruption, Prof. Huxley, on the contrary, took a seat on the London School Board !

We naturally think of Tyndall as a great scientific man. But he must also take a high place in the ranks of literature His descriptions of Alpine scenery especially contain many passages of vivid description and remarkable literary beauty. Take that in which he describes (in his account of the "Ascent of the Finsteraarhorn" in 1858) an Alpine valley in the Oberland :—

"I often turned," he says, "to look along this magnificent corridor. The mightiest mountains in the Oberland form its sides ; still, the impression which it makes is not that of vastness or sublimity, but of loveliness not to be described. The sun had not yet smitten the snows of the bounding mountains, but the saddle carved out a segment of the heavens which formed a background of unspeakable beauty. Over the rim of the saddle the sky was deep orange, passing upwards through amber, yellow, and vague ethereal green to the ordinary firmamental blue. Right above the snow-curve purple clouds hung perfectly motionless, giving depth to the spaces between them. There was something saintly in the scene. Anything more exquisite I had never beheld. We marched upwards over the smooth crisp snow to the crest of the saddle, and here I turned to take a last look along that grand corridor, and at that wonderful 'daffodil sky.' The sun's rays had already smitten the snows of the Aletschhorn ; the radiance seemed to infuse a principle of life and activity into the mountains and glaciers, but still that holy light shone forth, and those motionless clouds floated beyond, reminding one of that eastern religion whose essence is the repression of all action and the substitution for it of immortal calm."

Or take, in a different vein, that fine passage from Mountaineering in 1861 :—

"Skies and summits are to-day without a cloud, and no mist or turbidity interferes with the sharpness of the outlines. Jungfrau, Monk, Eiger, Trugberg, cliffy Strahlgrat, stately ladylike Aletschhorn, all grandly pierce the empyrean. Like a Saul of Mountains, the Finsteraarhorn overtops all his neighbours ; then we have the Oberaarhorn, with the riven glacier of Viesch rolling from his shoulders. Below is the Mârjelin See, with its crystal precipices and its floating icebergs, snowy white, sailing on a blue-green sea. Beyond is the range which divides the Valais from Italy. Sweeping round, the vision meets an aggregate of peaks which look as fledglings to their mother towards the mighty Dom. Then come the repellent crags of Mont Cervin; the ideal of moral savagery, of wild untamable ferocity, mingling involuntarily with our contemplation of the gloomy pile. Next comes an object, scarcely less grand, conveying, it may be, even a deeper impression of majesty and might than the Matterhorn itself—the Weisshorn, perhaps the most splendid object in the Alps. But beauty is associated with its force, and we think of it, not as cruel, but as grand and strong. Further to the right the great Combin lifts up his bare head; other peaks crowd around him; while at the extremity of the curve round which our gaze has swept rises the sovran crown of Mont Blanc. And now, as day sinks, scrolls of pearly clouds draw themselves around the mountain crests, being wafted from them into the distant air. They are without colour of any kind; still, by grace of form, and as the embodiment of lustrous light and most tender shade, their beauty is not to be described."

Among those who have described the splendid natural phenomena of the Alps and of the mountain heights with both imagination and science, Tyndall stands in the front rank; and I am glad to think that his writings in this kind are not likely to be forgotten by the new generation.

1906.

THE following is a list of Tyndall's published works:—"On the Magneto-optic Properties of Crystals and the Relation of Magnetism and Diamagnetism to Molecular Arrangement" (with Knoblauch), *Phil. Mag.*, 1850. "Phenomena of a Water Jet" (*Phil. Mag.*), 1851. "On the Importance of the Study of Physics" (Royal Institution), 1855. "A Day among the Séracs of the Glacier du Géant," 1859. "The Glaciers of the Alps," 1860. "Mountaineering in 1861, 1862." "Lecture on Experimental Physics," April 30, 1861. "Heat considered as a Mode of Motion" (Twelve Lectures, Royal Institution), 1863, 1865, &c.; with additions. "The Constitution of the Universe" (*Fortnightly Review*, December 1, 1865). "On Radiation" ("Rede" Lecture), 1865. "On Sound" (Eight Lectures, Royal Institution), 1867, 1875, 1883, 1893. "Miracles and Special Providences" (*Fortnightly Review*, June 1, 1867). "Faraday as a Discoverer," 1868. "Natural Philosophy, in Easy Lessons," 1869. "Notes of a Course of Nine Lectures on Light" (Royal Institution), 1870. "On the Scientific Use of the Imagination" (Brit. Ass.), 1870. "Researches on Diamagnetism and Magne-crystallic Action, &c.," 1870. "Notes on Electrical Phenomena and Theories" (Seven Lectures), 1870. "Hours of Exercise in the Alps," 1871, 1873. "Fragments of Science for Unscientific People," 1871; enlarged in following editions. "Contributions to Molecular Physics in the Domain of Radiant Heat," 1872. "The Forms of Water in Clouds, Rivers, Ice and Glaciers," 1872. "Six Lectures on Light," 1873. "Principal Forbes and his Biographers," 1873. "Addresses delivered before the British Association at Belfast," 1874. "Lessons in Electricity at the Royal Institution," 1875-6, 1876. "Fermentation and its Bearings on the Phenomena of Disease," 1877. "The Sabbath: Presidential Address to Glasgow Sunday Society," 1880. "Essays on the Floating Matter of the Air," 1881. "Lighthouse Illuminants" (Letters to *Times*, 1885). "New Fragments," 1892, 1897.

TO

MICHAEL FARADAY

THIS BOOK

IS AFFECTIONATELY INSCRIBED

1860

GLACIERS OF THE ALPS

INTRODUCTORY

I

IN the autumn of 1854 I attended the meeting of the British Association at Liverpool; and, after it was over, availed myself of my position to make an excursion into North Wales. Guided by a friend who knew the country, I became acquainted with its chief beauties, and concluded the expedition by a visit to Bangor and the neighbouring slate quarries of Penrhyn.

From my boyhood I had been accustomed to handle slates; had seen them used as roofing materials, and had worked the usual amount of arithmetic upon them at school; but now, as I saw the rocks blasted, the broken masses removed to the sheds surrounding the quarry, and there cloven into thin plates, a new interest was excited, and I could not help asking after the cause of this extraordinary property of cleavage. It sufficed to strike the point of an iron instrument into the edge of a plate of rock to cause the mass to yield and open, as wood opens in advance of a wedge driven into it. I walked round the quarry and observed that the planes of cleavage were everywhere parallel; the rock was capable of being split in one direction only, and this direction remained perfectly constant throughout the entire quarry.

I was puzzled, and, on expressing my perplexity to my companion, he suggested that the cleavage was nothing more than the layers in which the rock had been originally deposited, and which, by some subsequent disturbance, had been set on end, like the strata of the sandstone rocks and chalk cliffs of Alum Bay. But though I was too ignorant to combat this notion successfully, it by no means satisfied me. I did not know that at the time of my visit this very question of slaty cleavage was exciting the greatest attention among English geologists, and I quitted the place with that feeling of intellectual discontent which, however unpleasant it may be for a time, is very useful as a stimulant, and perhaps as necessary to the true appreciation of knowledge as a healthy appetite is to the enjoyment of food.

On inquiry I found that the subject had been treated by three English writers, Professor Sedgwick, Mr. Daniel Sharpe, and Mr. Sorby. From Professor Sedgwick I learned that cleavage and stratification were things totally distinct from each other; that in many cases the strata could be observed with the cleavage passing through them at a high angle; and that this was the case throughout vast areas in North Wales and Cumberland. I read the lucid and important memoir of this eminent geologist with great interest: it placed the data of the problem before me, as far as they were then known, and I found myself, to some extent at least, in a condition to appreciate the value of a theoretic explanation.

Everybody has heard of the force of gravitation, and of that of cohesion; but there is a more subtle play of forces exerted by the molecules of bodies upon each other when these molecules possess sufficient freedom of action. In virtue of such forces, the ultimate particles of matter are enabled to build themselves up into those wondrous edifices which we call crystals. A diamond is a crystal self-erected from atoms of carbon; an amethyst is a crystal built up from particles of silica; Iceland spar is a crystal built by particles of carbonate of lime. By artificial means we can allow the particles of bodies the free play necessary to their crystallisation. Thus a solution of saltpetre exposed to slow evaporation produces crystals of saltpetre; alum crystals of great size and beauty may be obtained in a similar manner; and in the formation of a bit of common sugar-candy there are agencies at play, the contemplation of which, as mere objects of thought, is sufficient to make the wisest philosopher bow down in wonder, and confess himself a child.

The particles of certain crystalline bodies are found to arrange themselves in layers, like courses of atomic masonry, and along these layers such crystals may be easily cloven into the thinnest laminæ. Some crystals possess *one* such direction in which they may be cloven, some several; some, on the other hand, may be split with different facility in different directions. Rock salt may be cloven with equal facility in three directions at right angles to each other; that is, it may be split into cubes; calcspar may be cloven in three directions oblique to each other; that is, into rhomboids. Heavy spar may also be cloven in three directions, but one cleavage is much more perfect, or more *eminent* as it is sometimes called, than the rest. Mica is a crystal which cleaves very readily in one direction, and it is sufficiently tough to furnish films of extreme tenuity: finally, any boy, with sufficient skill, who tries a good crystal of sugar-candy in various directions with the blade of his penknife, will find that it possesses one direction in particular, along which, if the blade of the knife be placed and struck, the crystal will split into plates possessing clean and shining surfaces of cleavage.

Professor Sedgwick was intimately acquainted with all these facts, and a great many more, when he investigated the cleavage of slate-rocks; and seeing no other explanation open to him, he ascribed to slaty cleavage a crystalline origin. He supposed that the particles of slate-rock were acted on, after their deposition, by "polar forces," which so arranged them as to produce the cleavage. According to this theory, therefore, Honister Crag and the cliffs of Penrhyn are to be regarded as portions of enormous crystals; a length of time commensurate with the vastness of the supposed action being assumed to have elapsed between the deposition of the rock and its final crystallisation.

When, however, we look closely into this bold and

beautiful hypothesis, we find that the only analogy which exists between the physical structure of slate-rocks and of crystals is this single one of cleavage. Such a coincidence might fairly give rise to the conjecture that both were due to a common cause; but there is great difficulty in accepting this as a theoretic truth. When we examine the structure of a slate-rock, we find that the substance is composed of the débris of former rocks; that it was once a fine mud, composed of particles of *sensible magnitude*. Is it meant that these particles, each taken as a whole, were rearranged after deposition? If so, the force which effected such an arrangement must be wholly different from that of crystallisation, for the latter is essentially *molecular*. What is this force? Nature, as far as we know, furnishes none competent, under the conditions, to produce the effect. Is it meant that the molecules composing these sensible particles have rearranged themselves? We find no evidence of such an action in the individual fragments: the mica is still mica, and possesses all the properties of mica; and so of the other ingredients of which the rock is composed. Independent of this, that an aggregate of heterogeneous mineral fragments should, without any assignable external cause, so shift its molecules as to produce a plane of cleavage common to them all, is, in my opinion, an assumption too heavy for any theory to bear.

Nevertheless, the paper of Professor Sedgwick invested the subject of slaty cleavage with an interest not to be forgotten, and proved the stimulus to further inquiry. The structure of slate-rocks was more closely examined; the fossils which they contained were subjected to rigid scrutiny, and their shapes compared with those of the same species taken from other rocks. Thus proceeding, the late Mr. Daniel Sharpe found that the fossils contained in slate-rocks are distorted in shape, being uniformly flattened out in the direction of the planes of cleavage. Here, then, was a fact of capital importance,— the shells became the indicators of an action to which the mass containing them had been subjected; they demonstrated the operation of pressure acting at right angles to the planes of cleavage.

The more the subject was investigated, the more clearly were the evidences of pressure made out. Subsequent to Mr. Sharpe, Mr. Sorby entered upon this field of inquiry. With great skill and patience he prepared sections of slate-rock, which he submitted to microscopic examination, and his observations showed that the evidences of pressure could be plainly traced, even in his minute specimens. The subject has been since ably followed up by Professors Haughton, Harkness, and others; but to the two gentlemen first mentioned we are, I think, indebted for the prime facts on which rests the *mechanical theory* of slaty cleavage.[1]

The observations just referred to showed the co-existence of the two phenomena, but they did not prove that pressure and cleavage stood to each other in the relation of cause and effect. "Can the pressure produce the cleavage?" was still an open question, and it was one which mere reasoning, unaided by experiment, was incompetent to answer.

Sharpe despaired of an experimental solution, regarding our means as inadequate, and our time on earth too short to produce the result. Mr. Sorby was more hopeful. Submitting mixtures of gypsum and oxide of iron scales to pressure, he found that the scales set themselves approximately at right angles to the direction in which the pressure was applied. The position of the scales resembled that of the plates of mica which his researches had disclosed to him in slate-rock, and he inferred that the presence of such plates, and of flat or elongated fragments generally, lying all in the same general direction, was the cause of slaty cleavage. At the meeting of the British Association at Glasgow, in 1855, I had the pleasure of seeing some of Mr. Sorby's specimens, and,

---

[1] Mr. Sorby has drawn my attention to an able and interesting paper by M. Bauer, in Karsten's 'Archiv' for 1846; in which it is announced that cleavage is a tension of the mass *produced by pressure*. The author refers to the experiments of Mr. Hopkins as bearing upon the question.

though the cleavage they exhibited was very rough, still, the tendency to yield at right angles to the direction in which the pressure had been applied, appeared sufficiently manifest.

At the time now referred to I was engaged, and had been for a long time previously, in examining the effects of pressure upon the magnetic force, and, as far back as 1851, I had noticed that some of the bodies which I had subjected to pressure exhibited a cleavage of surpassing beauty and delicacy. The bearing of such facts upon the present question now forcibly occurred to me. I followed up the observations; visited slate-yards and quarries, observed the exfoliation of rails, the fibres of iron, the structure of tiles, pottery, and cheese, and had several practical lessons in the manufacture of puff-paste and other laminated confectionery. My observations, I thought, pointed to a theory of slaty cleavage different from any previously given, and which, moreover, referred a great number of apparently unrelated phenomena to a common cause. On the 10th of June 1856 I made them the subject of a Friday evening's discourse at the Royal Institution.

Such are the circumstances, apparently remote enough, under which my connection with glaciers originated. My friend, Professor Huxley, was present at the lecture referred to: he was well acquainted with the work of Professor Forbes, entitled 'Travels in the Alps,' and he surmised that the question of slaty cleavage, in its new aspect, might have some bearing upon the laminated structure of glacier-ice discussed in the work referred to. He therefore urged me to read the 'Travels,' which I did with care, and the book made the same impression upon me that it had produced upon my friend. We were both going to Switzerland that year, and it required but a slight modification of our plans to arrange a joint excursion over some of the glaciers of the Oberland, and thus afford ourselves the means of observing together the veined structure of the ice.

Had the results of this arrangement been revealed to me beforehand, I should have paused before entering

Introductory

upon an investigation which required of me so long a renunciation of my old and more favourite pursuits. But no man knows when he commences the examination of a physical problem into what new and complicated mental alliances it may lead him. No fragment of nature can be studied alone: each part is related to every other part; and hence it is, that, following up the links of law which connect phenomena, the physical investigator often finds himself led far beyond the scope of his original intentions, the danger in this respect augmenting in direct proportion to the wish of the inquirer to render his knowledge solid and complete.

When the idea of writing this book first occurred to me, it was not my intention to confine myself to the glaciers alone, but to make the work a vehicle for the familiar explanation of such general physical phenomena as had come under my notice. Nor did I intend to address it to a cultured man of science, but to a *youth* of average intelligence, and furnished with the education which England now offers to the young. I wished indeed to make it a boy's class-book, which should reveal the mode of life, as well as the scientific objects, of an explorer of the Alps. The incidents of the past year have caused me to deviate, in some degree, from this intention, but its traces will be sufficiently manifest; and this reference to it will, I trust, excuse an occasional liberty of style and simplicity of treatment which would be out of place if intended for a reader of riper years.

EXPEDITION OF 1856

THE OBERLAND

II

On the 16th of August 1856 I received my alpenstock from the hands of Dr. Hooker, in the garden of the Pension Ober, at Interlaken. It bore my name, not marked, however, by the vulgar brands of the country, but by the solar beams which had been converged upon it by the pocket lens of my friend. I was the companion of Mr. Huxley, and our first aim was to cross the Wengern Alp. Light and shadow enriched the crags and green slopes as we advanced up the valley of Lauterbrunnen, and each occupied himself with that which most interested him. My companion examined the drift, I the cleavage, while both of us looked with interest at the contortions of the strata to our left, and at the shadowy, unsubstantial aspect of the pines, gleaming through the sunhaze to our right.

What was the physical condition of the rock when it was thus bent and folded like a pliant mass? Was it necessarily softer than it is at present? I do not think so. The shock which would crush a railway carriage, if communicated to it at once, is harmless when distributed over the interval necessary for the pushing in of the buffer. By suddenly stopping a cock from which water flows you may burst the conveyance pipe, while a slow turning of the cock keeps all safe. Might not a solid rock by ages of pressure be folded as above? It is a physical axiom that no body is perfectly hard, none perfectly soft, none perfectly elastic. The hardest body subjected to pressure yields, however little, and the same body when the pressure is removed *cannot return* to its original form. If it did not yield in the slightest degree it would be perfectly

The Oberland

hard; if it could completely return to its original shape it would be perfectly elastic.

Let a pound weight be placed upon a cube of granite; the cube is flattened, though in an infinitesimal degree. Let the weight be removed, the cube *remains* a little flattened; it cannot quite return to its primitive condition. Let us call the cube thus flattened No. 1. Starting with No. 1 as a new mass, let the pound weight be laid upon it; the mass yields, and on removing the weight it cannot return to the dimensions of No. 1; we have a more flattened mass, No. 2. Proceeding in this manner, it is manifest that by a repetition of the process we should produce a series of masses, each succeeding one more flattened than the former. This appears to be a necessary consequence of the physical axiom referred to above.

Now if, instead of removing and replacing the weight in the manner supposed, we cause it to rest continuously upon the cube, the flattening, which above was intermittent, will be continuous; no matter how hard the cube may be, there will be a gradual yielding of its mass under the pressure. Apply this to squeezed rocks—to those, for example, which form the base of an obelisk like the Matterhorn; that this base must yield, seems a certain consequence of the physical constitution of matter: the conclusion seems inevitable that the mountain is sinking by its own weight. Let two points be fixed, one near the summit, the other near the base of the obelisk; next year these points will have approached each other. Whether the amount of approach in a human lifetime be measurable we know not; but it seems certain that ages would leave their impress upon the mass, and render visible to the eye an action which at present is appreciable by the imagination only.

We halted on the night of the 16th at the Jungfrau Hotel, and next morning we saw the beams of the rising sun fall upon the peaked snow of the Silberhorn. Slowly and solemnly the pure white cone appeared to rise higher and higher into the sunlight, being afterwards mottled with gold and gloom, as clouds drifted between it and the sun. I descended alone towards the base of the mountain, making my way through a rugged gorge, the sides of which were strewn with pine-trees, splintered, broken across, and torn up by the roots. I finally reached the end of a glacier, formed by the snow and shattered ice which fall from the shoulders of the Jungfrau. The view from this place had a savage magnificence such as I had not previously beheld, and it was not without some slight feeling of awe that I clambered up the end of the glacier. It was the first I had actually stood upon. The loneliness of the place was very impressive, the silence being only broken by fitful gusts of wind, or by the weird rattle of the débris which fell at intervals from the melting ice.

Once I noticed what appeared to be the sudden and enormous augmentation of the waters of a cascade, but the sound soon informed me that the increase was due to an avalanche which had chosen the track of the cascade for its rush. Soon afterwards my eyes were fixed upon a white slope some thousands of feet above me; I saw the ice give way, and, after a sensible interval, the thunder of another avalanche reached me. A kind of zigzag channel had been worn on the side of the mountain, and through this the avalanche rushed, hidden at intervals, and anon shooting forth, and leaping like a cataract down the precipices. The sound was sometimes continuous, but sometimes broken into rounded explosions which seemed to assert a passionate predominance over the general level of the roar. These avalanches, when they first give way, usually consist of enormous blocks of ice, which are more and more shattered as they descend. Partly to the echoes of the first crash, but mainly, I think, to the shock of the harder masses which preserve their cohesion, the explosions which occur during the descent of the avalanche are to be ascribed. Much of the ice is crushed to powder; and thus, when an avalanche pours cataract-like over a ledge, the heavier masses, being less influenced by the atmospheric resistance, shoot forward like descending rockets, leaving the lighter powder in trains behind them. Such is the material of which a class of the smaller glaciers in the Alps is composed.

The Oberland

They are the products of avalanches, the crushed ice being recompacted into a solid mass, which exhibits on a smaller scale most of the characteristics of the large glaciers.

After three hours' absence I reascended to the hotel, breakfasted, and afterwards returned with Mr. Huxley to the glacier. While we were engaged upon it the weather suddenly changed; lightning flashed about the summits of the Jungfrau, and thunder "leaped" among her crags. Heavy rain fell, but it cleared up afterwards with magical speed, and we returned to our hotel. Heedless of the forebodings of many prophets of evil weather we set out for Grindelwald. The scene from the summit of the little Scheideck was exceedingly grand. The upper air exhibited a commotion which we did not experience; clouds were wildly driven against the flanks of the Eiger, the Jungfrau thundered behind, while in front of us a magnificent rainbow, fixing one of its arms in the valley of Grindelwald, and, throwing the other right over the crown of the Wetterhorn, clasped the mountain in its embrace. Through jagged apertures in the clouds floods of golden light were poured down the sides of the mountain. On the slopes were innumerable châlets, glistening in the sunbeams, herds browsing peacefully and shaking their mellow bells; while the blackness of the pine-trees, crowded into woods, or scattered in pleasant clusters over alp and valley, contrasted forcibly with the lively green of the fields.

At Grindelwald, on the 18th, we engaged a strong and competent guide, named Christian Kaufmann, and proceeded to the Lower Glacier. After a steep ascent, we gained a point from which we could look down upon the frozen mass. At first the ice presented an appearance of utter confusion, but we soon reached a position where the mechanical conditions of the glacier revealed themselves, and where we might learn, had we not known it before, that confusion is merely the unknown intermixture of laws, and becomes order and beauty when we rise to their comprehension. We reached the so-called Eismeer—Ice Sea. In front of us was the range of the Viescherhörner, and a vast snow slope, from which one branch of the glacier was fed. Near the base of this *névé*, and surrounded on all sides by ice, lay a brown rock, to which our attention was directed as a place noted for avalanches; on this rock snow or ice never rests, and it is hence called the *Heisse Platte*—the Hot Plate. At the base of the rock, and far below it, the glacier was covered with clean crushed ice, which had fallen from a crown of frozen cliffs

FIG. 1

encircling the brow of the rock. One obelisk in particular signalised itself from all others by its exceeding grace and beauty. Its general surface was dazzling white, but from its clefts and fissures issued a delicate blue light, which deepened in hue from the edges inwards. It stood upon a pedestal of its own substance, and seemed as accurately fixed as if rule and plummet had been employed in its erection. Fig. 1 represents this beautiful minaret of ice.

While we were in sight of the *Heisse Platte*, a dozen avalanches rushed downwards from its summit. In most cases we were informed of the descent of an avalanche by the sound, but sometimes the white mass was seen gliding down the rock, and scattering its *smoke* in the air, long before the sound reached us. It is difficult to reconcile the insignificant appearance presented by avalanches, when seen from a distance, with the volume of sound which they generate; but on this day we saw sufficient to account for the noise. One block of solid ice which we found below the *Heisse Platte* measured 7 feet 6 inches in length, 5 feet 8 inches in height, and 4 feet 6 inches in depth. A second mass was 10 feet long, 8 feet high, and 6 feet wide. It contained therefore 480 cubit feet of ice, which had been cast to a distance of nearly 1000 yards down the glacier. The shock of such hard and ponderous projectiles against rocks and ice, reinforced by the echoes from the surrounding mountains, will appear sufficient to account for the peals by which their descent is accompanied.

A second day, in company with Dr. Hooker, completed the examination of this glacier in 1856; after which I parted from my friends, Mr. Huxley intending to rejoin me at the Grimsel. On the morning of the 20th of August I strapped on my knapsack and ascended the green slopes from Grindelwald towards the great Scheideck. Before reaching the summit I frequently heard the wonderful echoes of the Wetterhorn. Some travellers were in advance of me, and to amuse *them* an alpine horn was blown. The direct sound was cut off from me by a hill, but the echoes talked down to me from the mountain walls. The sonorous waves arrived after one, two, three, and more reflections, diminishing gradually in intensity, but increasing in softness, as if in its wanderings from crag to crag the sound had undergone a kind of sifting process, leaving all its grossness behind, and returning in delightful flute notes to the ear.

Let us investigate this point a little. If two looking-glasses be placed perfectly parallel to each other, with a lighted candle between them, an infinite series of images of the candle will be seen at both sides, the images diminishing in brightness the further they recede. But if the looking-glasses, instead of being parallel, enclose an angle, a limited number of images only will be seen. The smaller the angle which the reflectors make with each other, or, in other words, the nearer they approach parallelism, the greater will be the number of images observed.

To find the number of images the following is the rule:—Divide 360, or the number of degrees in a circle, by the number of degrees in the angle enclosed by the two mirrors, the quotient will be *one more* than the number of images; or, counting the object itself, the quotient is always equal to the number of images plus the object. In Fig. 2 I have given the number and position of the images produced by two mirrors placed at an angle of 45°. A B and B C mark the edges of the mirrors, and O represents the candle, which, for the sake of simplicity, I have placed midway between them. Fix one point of a pair of compasses at B, and with the distance B O sweep a circle—*all the images will be ranged upon the circumference of this circle.* The number of images found by the foregoing rule is 7, and their positions are marked in the figure by the numbers 1, 2, 3, &c.

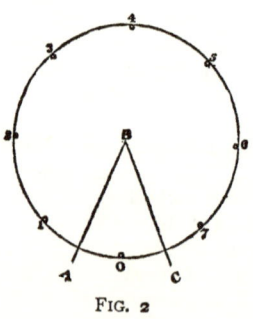
FIG. 2

Suppose the *ear* to occupy the place of the eye, and that *a sounding body* occupies the place of the luminous one, we should then have just as many *echoes* as we had *images* in the former case. These echoes would diminish in loudness just as the images of the candle diminish in brightness. At each reflection a portion both of sound and light is lost; hence the oftener light is reflected the dimmer it becomes, and the oftener sound is reflected the fainter it is.

Now the cliffs of the Wetterhorn are so many rough angular reflectors of the sound: some of them send it

The Oberland

back directly to the listener, and we have a first echo; some of them send it on to others from which it is again reflected, forming a second echo. Thus, by repeated reflection, successive echoes are sent to the ear, until, at length, they become so faint as to be inaudible. The sound, as it diminishes in intensity, appears to come from greater and greater distances, as if it were receding into the mountain solitudes; the final echoes being inexpressibly soft and pure.

After crossing the Scheideck I descended to Meyringen, visiting the Reichenbach waterfall on my way. A peculiarity of the descending water here is, that it is broken up in one of the basins into nodular masses, each of which in falling leaves the light foaming mass which surrounds it as a train in the air behind; the effect exactly resembles that of the avalanches of the Jungfrau, in which the more solid blocks of ice shoot forward in advance of the lighter débris, which is held back by the friction of the air.

Next day I ascended the valley of Hasli, and observed upon the rocks and mountains the action of ancient glaciers which once filled the valley to the height of more than a thousand feet above its present level. I paused, of course, at the waterfall of Handeck, and stood for a time upon the wooden bridge which spans the river at its top. The Aar comes gambolling down to the bridge from its parent glacier, takes one short jump upon a projecting ledge, boils up into foam, and then leaps into a chasm, from the bottom of which its roar ascends through the gloom. A rivulet named the Aarlenbach joins the Aar from the left in the very jaws of the chasm: falling, at first, upon a projection at some depth below the edge, and, rebounding from this, it darts at the Aar, and both plunge together like a pair of fighting demons to the bottom of the gorge. The foam of the Aarlenbach is white, that of the Aar is yellow, and this enables the observer to trace the passage of the one cataract *through* the other. As I stood upon the bridge the sun shone brightly upon the spray and foam; my shadow was oblique to the river, and hence a symmetrical rainbow could not be formed in the spray, but one half of a lovely

B

bow, with its base in the chasm, leaned over against the opposite rocks, the colours advancing and retreating as the spray shifted its position. I had been watching the water intently for some time, when a little Swiss boy, who stood beside me, observed, in his trenchant German, "There plunge stones ever downwards." The stones were palpable enough, carried down by the cataract, and sometimes completely breaking loose from it, but I did not see them until my attention was withdrawn from the water.

On my arrival at the Grimsel I found Mr. Huxley already there, and, after a few minutes' conversation, we decided to spend a night in a hut built by M. Dolfuss in 1846, beside the Unteraar Glacier, about 2000 feet above the Hospice. We hoped thus to be able to examine the glacier to its origin on the following day. Two days' food and some blankets were sent up from the Hospice, and, accompanied by our guide, we proceeded to the glacier.

Having climbed a great terminal moraine, and tramped for a considerable time amid loose shingle and boulders, we came upon the ice. The finest specimens of "tables" which I have ever seen are to be found upon this glacier —huge masses of clean granite poised on pedestals of ice. Here are also "dirt-cones" of the largest size, and numerous shafts, the forsaken passages of ancient "moulins," some filled with water, others simply with deep blue light. I reserve the description and explanation of both cones and moulins for another place. The surfaces of some of the small pools were sprinkled lightly over with snow, which the water underneath was unable to melt; a coating of snow granules was thus formed, flexible as chain armour, but so close that the air could not escape through it. Some bubbles which had risen through the water had lifted the coating here and there into little rounded domes, which, by gentle pressure, could be shifted hither and thither, and several of them collected into one. We reached the hut, the floor of which appeared to be of the original mountain slab; there was a space for cooking walled off from the sleeping-room, half of which was raised above the floor, and contained a quantity of old hay. The number 2404 mètres, the height, I suppose, of the place above the sea, was painted on the door, behind which were also the names of several well-known observers—Agassiz, Forbes, Desor, Dolfuss, Ramsay, and others—cut in the wood. A loft contained a number of instruments for boring, a surveyor's chain, ropes, and other matters. After dinner I made my way alone towards the junction of the Finsteraar and Lauteraar Glaciers, which unite at the Abschwung to form the trunk stream of the Unteraar Glacier. Upon the great central moraine which runs between the branches were perched enormous masses of rock, and, under the overhanging ledge of one of these, M. Agassiz had his *Hôtel des Neufchâtelois*. The rock is still there, bearing traces of names now nearly obliterated by the weather, while the fragments around also bear inscriptions. There in the wilderness, in the grey light of evening, these blurred and faded evidences of human activity wore an aspect of sadness. It was a temple of science now in ruins, and I a solitary pilgrim to the desecrated blocks. As the day declined, rain began to fall, and I turned my face towards my new home; where in due time we betook ourselves to our hay, and waited hopefully for the morning.

But our hopes were doomed to disappointment. A vast quantity of snow fell during the night, and, when we arose, we found the glacier covered, and the air thick with the descending flakes. We waited, hoping that it might clear up, but noon arrived and passed without improvement; our firewood was exhausted, the weather intensely cold, and, according to the men's opinion, hopelessly bad; they opposed the idea of ascending further, and we had therefore no alternative but to pack up and move downwards. What was snow at the higher elevations changed to rain lower down, and drenched us completely before we reached the Grimsel. But though thus partially foiled in our design, this visit taught us much regarding the structure and general phenomena of the glacier.

The morning of the 24th was clear and calm: we rose with the sun, refreshed and strong, and crossed the

Grimsel Pass at an early hour. The view from the summit of the pass was lovely in the extreme; the sky a deep blue, the surrounding summits all enamelled with the newly fallen snow, which gleamed with dazzling whiteness in the sunlight. It was Sunday, and the scene was itself a Sabbath, with no sound to disturb its perfect rest. In a lake which we passed the mountains were mirrored without distortion, for there was no motion of the air to ruffle its surface. From the summit of the Mayenwand we looked down upon the Rhone Glacier, and a noble object it seemed,—I hardly know a finer of its kind in the Alps. Forcing itself through the narrow gorge which holds the ice cascade in its jaws, and where it is greatly riven and dislocated, it spreads out in the valley below in such a manner as clearly to reveal to the mind's eye the nature of the forces to which it is subjected. Longfellow's figure is quite correct; the glacier resembles a vast gauntlet, of which the gorge represents the wrist; while the lower glacier, cleft by its fissures into finger-like ridges, is typified by the hand.

Furnishing ourselves with provisions at the adjacent inn, we devoted some hours to the examination of the lower portion of the glacier. The dirt upon its surface was arranged in grooves as fine as if produced by the passage of a rake, while the laminated structure of the deeper ice always corresponded to the superficial grooving. We found several shafts, some empty, some filled with water. At one place our attention was attracted by a singular noise, evidently produced by the forcing of air and water through passages in the body of the glacier; the sound rose and fell for several minutes, like a kind of intermittent snore, reminding one of Hugi's hypothesis that the glacier was alive.

We afterwards climbed to a point from which the whole glacier was visible to us from its origin to its end. Adjacent to us rose the mighty mass of the Finsteraarhorn, the monarch of the Oberland. The Galenstock was also at hand, while round about the *névé* of the glacier a mountain wall projected its jagged outline against the sky. At a distance was the grand cone of the Weisshorn, then, and I believe still, unscaled; further to the left the magnificent peaks of the Mischabel; while between them, in savage isolation, stood the obelisk of the Matterhorn. Near us was the chain of the Furca, all covered with shining snow, while overhead the dark blue of the firmament so influenced the general scene as to inspire a sentiment of wonder approaching to awe. We descended to the glacier, and proceeded towards its source. As we advanced an unusual light fell upon the mountains, and looking upwards we saw a series of coloured rings, drawn like a vivid circular rainbow quite round the sun. Between the orb and us spread a thin veil of cloud on which the circles were painted; the western side of the veil soon melted away, and with it the colours, but the eastern half remained a quarter of an hour longer, and then in its turn disappeared. The crevasses became more frequent and dangerous as we ascended. They were usually furnished with overhanging eaves of snow, from which long icicles depended, and to tread on which might be fatal. We were near the source of the glacier, but the time necessary to reach it was nevertheless indefinite, so great was the entanglement of fissures. We followed one huge chasm for some hundreds of yards, hoping to cross it; but after half-an-hour's fruitless effort we found ourselves baffled and forced to retrace our steps.

The sun was sloping to the west, and we thought it wise to return; so down the glacier we went, mingling our footsteps with the tracks of chamois, while the frightened marmots piped incessantly from the rocks. We reached the land once more, and halted for a time to look upon the scene within view. The marvellous blueness of the sky in the earlier part of the day indicated that the air was charged, almost to saturation, with transparent aqueous vapour. As the sun sank the shadow of the Finsteraarhorn was cast through the adjacent atmosphere, which, thus deprived of the direct rays, curdled up into visible fog. The condensed vapour moved slowly along the flanks of the mountain, and poured itself cataract-like into the valley of the Rhone.

Here it met the sun again, which reduced it once more to the invisible state. Thus, though there was an incessant supply from the generator behind, the fog made no progress; as in the case of the moving glacier, the end of the cloud-river remained stationary where consumption was equal to supply. Proceeding along the mountain to the Furca, we found the valley at the further side of the pass also filled with fog, which rose, like a wall, high above the region of actual shadow. Once on turning a corner an exclamation of surprise burst simultaneously from my companion and myself. Before each of us and against the wall of fog, stood a spectral image of a man, of colossal dimensions; dark as a whole, but bounded by a coloured outline. We stretched forth our arms; the spectres did the same. We raised our alpenstocks; the spectres also flourished their batons. All our actions were imitated by these fringed and gigantic shades. We had, in fact, *the Spirit of the Brocken* before us in perfection.

At the time here referred to I had had but little experience of alpine phenomena. I had been through the Oberland in 1850, but was then too ignorant to learn much from my excursion. Hence the novelty of this day's experience may have rendered it impressive: still even now I think there was an intrinsic grandeur in its phenomena which entitles the day to rank with the most remarkable that I have spent among the Alps. At the Furca, to my great regret, the joint ramblings of my friend and myself ended; I parted from him on the mountain side, and watched him descending, till the grey of evening finally hid him from my view.

THE TYROL

III

My subsequent destination was Vienna; but I wished to associate with my journey thither a visit to some of the glaciers of the Tyrol. At Landeck, on the 29th

The Tyrol

of August, I learned that the nearest glacier was that adjacent to the Gebatsch Alp, at the head of the Kaunserthal; and on the following morning I was on my way towards this valley. I sought to obtain a guide at Kaltebrunnen, but failed; and afterwards walked to the little hamlet of Feuchten, where I put up at a very lowly inn. My host, I believe, had never seen an Englishman, but he had heard of such, and remarked to me in his patois with emphasis, "Die Engländer sind die kühnste Leute in dieser Welt." Through his mediation I secured a chamois-hunter, named Johann Auer, to be my guide, and next morning I started with this man up the valley. The sun, as we ascended, smote the earth and us with great power; high mountains flanked us on either side, while in front of us, closing the view, was the mass of the Weisskugel, covered with snow. At three o'clock we came in sight of the glacier, and soon afterwards I made the acquaintance of the *Senner* or cheesemakers of the Gebatsch Alp.

The chief of these was a fine tall fellow, with free, frank countenance, which, however, had a dash of the mountain wildness in it. His feet were bare, he wore breeches, and fragments of stockings partially covered his legs, leaving a black zone between the upper rim of the sock and the breeches. His feet and face were of the same swarthy hue; still he was handsome, and in a measure pleasant to look upon. He asked me what he could cook for me, and I requested some bread and milk; the former was a month old, the latter was fresh and delicious, and on these I fared sumptuously. I went to the glacier afterwards with my guide, and remained upon the ice until twilight, when we returned, guided by no path, but passing amid crags grasped by the gnarled roots of the pine, through green dells, and over bilberry knolls of exquisite colouring. My guide kept in advance of me singing a Tyrolese melody, and his song and the surrounding scene revived and realised all the impressions of my boyhood regarding the Tyrol.

Milking was over when we returned to the châlet, which now contained four men exclusive of myself and my guide. A fire of pine logs was made upon a platform of stone, elevated three feet above the floor; there was no chimney, as the smoke found ample vent through the holes and fissures in the sides and roof. The men were all intensely sunburnt, the legitimate brown deepening into black with beard and dirt. The chief senner prepared supper, breaking eggs into a dish, and using his black fingers to empty the shell when the albumen was refractory. A fine erect figure he was as he stood in the glowing light of the fire. All the men were smoking, and now and then a brand was taken from the fire to light a renewed pipe, and a ruddy glare flung thereby over the wild countenance of the smoker. In one corner of the châlet, and raised high above the ground, was a large bed, covered with clothes of the most dubious black-brown hue; at one end was a little water-wheel turned by a brook, which communicated motion to a churndash which made the butter. The beams and rafters were covered with cheeses, drying in the warm smoke. The senner, at my request, showed me his storeroom, and explained to me the process of making cheese, its interest to me consisting in its bearing upon the question of slaty cleavage. Three gigantic masses of butter were in the room, and I amused my host by calling them butter-glaciers. Soon afterwards a bit of cotton was stuck in a lump of grease, which was placed in a lantern, and the wick ignited; the chamois-hunter took it, and led the way to our resting-place, I having previously declined a good-natured invitation to sleep in the big black bed already referred to.

There was a cowhouse near the châlet, and above it, raised on pillars of pine, and approached by a ladder, was a loft, which contained a quantity of dry hay: this my guide shook to soften the lumps, and erected an eminence for my head. I lay down, drawing my plaid over me, but Auer affirmed that this would not be a sufficient protection against the cold; he therefore piled hay upon me to the shoulders, and proposed covering up my head also. This, however, I declined, though the biting coldness of the air, which sometimes blew in upon us, afterwards proved to me the wisdom of the suggestion. Having set me right, my chamois-hunter prepared a place for himself, and soon his heavy breathing informed me that he was in a state of bliss which I could only envy. One by one the stars crossed the apertures in the roof. Once the Pleiades hung above me like a cluster of gems; I tried to admire them, but there was no fervour in my admiration. Sometimes I dozed, but always as this was about to deepen into positive sleep it was rudely broken by the clamour of a group of pigs which occupied the ground-floor of our dwelling. The object of each individual of the group was to secure for himself the maximum amount of heat, and hence the outside members were incessantly trying to become inside ones. It was the struggle of radical and conservative among the pachyderms, the politics being determined by the accident of position.

I rose at five o'clock on the 1st of September, and after a breakfast of black bread and milk ascended the glacier as far as practicable. We once quitted it, crossed a promontory, and descended upon one of its branches, which was flanked by some fine old moraines. We here came upon a group of seven marmots, which with yells of terror scattered themselves among the rocks. The points of the glacier beyond my reach I examined through a telescope; along the faces of the sections the lines of stratification were clearly shown; and in many places where the mass showed manifest signs of lateral pressure, I thought I could observe the cleavage passing through the strata. The point, however, was too important to rest upon an observation made from such a distance, and I therefore abstained from mentioning it subsequently. I examined the fissures and the veining, and noticed how the latter became most perfect in places where the pressure was greatest. The effect of *oblique* pressure was also finely shown: at one place the thrust of the descending glacier was opposed by the resistance offered by the side of the valley, the direction of the force being oblique to the side; the consequence was a structure nearly parallel to the valley, and consequently oblique to the thrust which I believe to be its cause.

After five hours' examination we returned to our châlet, where we refreshed ourselves, put our things in order, and faced a nameless "Joch," or pass; our aim being to cross the mountains into the valley of Lantaufer, and reach Graun that evening. After a rough ascent over the alp we came to the dead crag, where the weather had broken up the mountains into ruinous heaps of rock and shingle. We reached the end of a glacier, the ice of which was covered by sloppy snow, and at some distance up it came upon an islet of stones and débris, where we paused to rest ourselves. My guide, as usual, ranged over the summits with his telescope, and at length exclaimed, " I see a chamois." The creature stood upon a cliff some hundreds of yards to our left, and seemed to watch our movements. It was a most graceful animal, and its life and beauty stood out in forcible antithesis to the surrounding savagery and death.

On the steep slopes of the glacier I was assisted by the hand of my guide. In fact, on this day I deemed places dangerous, and dreaded them as such, which subsequent practice enabled me to regard with perfect indifference; so much does what we call courage depend upon habit, or on the fact of knowing that we have really nothing to fear. Doubtless there are times when a climber has to make up his mind for very unpleasant possibilities, and even gather calmness from the contemplation of the worst; but in most cases I should say that his courage is derived from the latent feeling that the chances of safety are immensely in his favour.

After a tough struggle we reached the narrow row of crags which form the crest of the pass, and looked into the world of mountain and cloud on the other side. The scene was one of stern grandeur—the misty lights and deep cloud-glooms being so disposed as to augment the impression of vastness which the scene conveyed. The breeze at the summit was exceedingly keen, but it gave our muscles tone, and we sprang swiftly downward through the yielding débris which here overlies the mountain, and in which we sometimes sank to the knees. Lower down we came once more upon the ice. The

The Tyrol

glacier had at one place melted away from its bounding cliff, which rose vertically to our right, while a wall of ice 60 or 80 feet high was on our left. Between both was a narrow passage, the floor of which was snow, which I knew to be hollow beneath: my companion, however, was in advance of me, and he being the heavier man, where he trod I followed without hesitation. On turning an angle of the rock I noticed an expression of concern upon his countenance, and he muttered audibly, " I did not expect this." The snow-floor had, in fact, given way, and exposed to view a clear green lake, one boundary of which was a sheer precipice of rock, and the other the aforesaid wall of ice; the latter, however, curved a little at its base, so as to form a short steep slope which overhung the water. My guide first tried the slope alone; biting the ice with his shoe nails, and holding on by the spike of his baton, he reached the other side. He then returned, and, divesting myself of all superfluous clothes, as a preparation for the plunge which I fully expected, I also passed in safety. Probably the consciousness that I had water to fall into instead of pure space, enabled me to get across without anxiety or mischance; but had I, like my guide, been unable to swim, my feelings would have been far different.

This accomplished, we went swiftly down the valley, and the more I saw of my guide the more I liked him. He might, if he wished, have made his day's journey shorter by stopping before he reached Graun, but he would not do so. Every word he said to me regarding distances was true, and there was not the slightest desire shown to magnify his own labour. I learnt by mere accident that the day's work had cut up his feet, but his cheerfulness and energy did not bate a jot till he had landed me in the Black Eagle at Graun. Next morning he came to my room, and said that he felt sufficiently refreshed to return home. I paid him what I owed him, when he took my hand, and, silently bending down his head, kissed it; then, standing erect, he stretched forth his right hand, which I grasped firmly in mine, and bade him farewell; and thus I parted from Johann Auer, my brave and truthful chamois-hunter.

On the following day I met Dr. Frankland in the Finstermuntz Pass, and that night we bivouacked together at Mals. Heavy rain fell throughout the night, but it came from a region high above that of liquidity. It was first snow, which, as it descended through the warmer strata of the atmosphere, was reduced to water. Overhead, in the air, might be traced a surface, below which the precipitate was liquid, above which it was solid; and this surface, intersecting the mountains which surround Mals, marked upon them a beautifully defined *snow-line*, below which the pines were dark and the pastures green, but above which pines and pastures and crags were covered with the freshly fallen snow.

On the 2nd of September we crossed the Stelvio. The brown cone of the well-known Madatschspitze was clear, but the higher summits were clouded, and the fragments of sunshine which reached the lower world wandered like gleams of fluorescent light over the glaciers. Near the snow-line the partial melting of the snow had rendered it coarsely granular, but as we ascended it became finer, and the light emitted from its cracks and cavities a pure and deep blue. When a staff was driven into the snow low down the mountain, the colour of the light in the orifice was scarcely sensibly blue, but higher up this increased in a wonderful degree, and at the summit the effect was marvellous. I struck my staff into the snow, and turned it round and round; the surrounding snow cracked repeatedly, and flashes of blue light issued from the fissures. The fragments of snow that adhered to the staff were, by contrast, of a beautiful pink yellow, so that, on moving the staff with such fragments attached to it up and down, it was difficult to resist the impression that a pink flame was ascending and descending in the hole. As we went down the other side of the pass, the effect became more and more feeble, until, near the snow-line, it almost wholly disappeared.

We remained that night at the baths of Bormio, but, the following afternoon being fine, we wished to avail ourselves of the fair weather to witness the scene from the summit of the pass. Twilight came on before we reached Santa Maria, but a gorgeous orange overspread the western horizon, from which we hoped to derive sufficient light. It was a little too late when we reached the top, but still the scene was magnificent. A multitude of mountains raised their crowns towards heaven, while above all rose the snow-white cone of the Ortler. Far into the valley the giant stretched his granite limbs, until they were hid from us by darkness. As this deepened, the heavens became more and more crowded with stars, which blazed like gems over the heads of the mountains. At times the silence was perfect, unbroken save by the crackling of the frozen snow beneath our own feet; while at other times a breeze would swoop down upon us, keen and hostile, scattering the snow from the roofs of the wooden galleries in frozen powder over us. Long after night had set in, a ghastly gleam rested upon the summit of the Ortler, while the peaks in front deepened to a dusky neutral tint, the more distant ones being lost in gloom. We descended at a swift pace to Trafoi, which we reached before 11 P.M.

Meran was our next resting-place, whence we turned through the Schnalzerthal to Unserfrau, and thence over the Hochjoch to Fend. From a religious procession we took a guide, who, though partly intoxicated, did his duty well. Before reaching the summit of the pass we were assailed by a violent hailstorm, each hailstone being a frozen cone with a rounded end. Had not their motion through the air something to do with the shape of these hailstones? The theory of meteorites now generally accepted is that they are small planetary bodies drawn to the earth by gravity, and brought to incandescence by friction against the earth's atmosphere. Such a body moving through the atmosphere must have condensed hot air in front of it, and rarefied cool air behind it; and the same is true to a small extent of a hailstone. This distribution of temperature must, I imagine, have some influence on the shape of the stone. Possibly also the stratified appearance of some hailstones may be connected with this action.[1]

The hail ceased and the heights above us cleared as we ascended. At the top of the pass we found ourselves on the verge of a great *névé*, which lay between two ranges of summits, sloping down to the base of each range from a high and rounded centre: a wilder glacier scene I have scarcely witnessed. Wishing to obtain a more perfect view of the region, I diverged from the track followed by Dr. Frankland and the guide, and climbed a ridge of snow about half a mile to the right of them. A glorious expanse was before me, stretching itself in vast undulations and heaping itself here and there into mountainous cones, white and pure, with the deep blue heaven behind them. Here I had my first experience of hidden crevasses, and to my extreme astonishment once found myself in the jaws of a fissure of whose existence I had not the slightest notice. Such accidents have often occurred to me since, but the impression made by the first is likely to remain the strongest. It was dark when we reached the wretched Wirthshaus at Fend, where, badly fed, badly lodged, and disturbed by the noise of innumerable rats, we spent the night. Thus ended my brief glacier expedition of 1856; and on the observations then made, and on subsequent experiments, was founded a paper presented to the Royal Society by Mr. Huxley and myself.

---

[1] I take the following account of a grander storm of the above character from Hooker's 'Himalayan Journals,' vol. ii. p. 405.

"On the 20th (March 1849) we had a change in the weather: a violent storm from the south-west occurred at noon, with hail of a strange form, the stones being sections of hollow spheres, half an inch across and upwards, formed of cones with truncated apices and convex bases: these cones were aggregated together with their bases outwards. The large masses were followed by a shower of the separate conical pieces, and that by heavy rain. On the mountains this storm was most severe: the stones lay at Dorjiling for seven days, congealed into masses of ice several feet long and a foot thick in sheltered places: at Purneah, fifty miles south, stones one and two inches across fell, probably as whole spheres."

EXPEDITION OF 1857

THE LAKE OF GENEVA

IV

THE time occupied in the observations of 1856 embraced about five whole days; and though these days were laborious and instructive, still so short a time proved to be wholly incommensurate with the claims of so wide a problem. During the subsequent experimental treatment of the subject, I had often occasion to feel the incompleteness of my knowledge, and hence arose the desire to make a second expedition to the Alps, for the purpose of expanding, fortifying, or, if necessary, correcting first impressions.

On Thursday, the 9th of July 1857, I found myself upon the Lake of Geneva, proceeding towards Vevey. I had long wished to see the waters of this renowned inland sea, the colour of which is perhaps more interesting to the man of science than to the poets who have sung about it. Long ago its depth of blue excited attention, but no systematic examination of the subject has, so far as I know, been attempted. It may be that the lake simply exhibits the colour of pure water. Ice is blue, and it is reasonable to suppose that the liquid obtained from the fusion of ice is of the same colour; but still the question presses—" Is the blue of the Lake of Geneva to be entirely accounted for in this way?" The attempts which have been made to explain it otherwise show that at least a doubt exists as to the sufficiency of the above explanation.

It is only in its deeper portions that the colour of the lake is properly seen. Where the bottom comes into view the pure effect of the water is disturbed; but where the water is deep the colour is deep: between Rolle and

Nyon, for example, the blue is superb. Where the blue was deepest, however, it gave me the impression of turbidity rather than of deep transparency. At the upper portion of the lake the water through which the steamer passed was of a blue green. Wishing to see the place where the Rhone enters the lake, I walked on the morning of the 10th from Villeneuve to Novelle, and thence through the woods to the river side. Proceeding along an embankment, raised to defend the adjacent land from the incursions of the river, an hour brought me to the place where it empties itself into the lake. The contrast between the two waters was very great: the river was almost white with the finely divided matter which it held in suspension; while the lake at some distance was of a deep ultramarine.

The lake in fact forms a reservoir where the particles held in suspension by the river have time to subside, and its waters to become pure. The subsidence of course takes place most copiously at the head of the lake; and here the deposit continues to form new land, adding year by year to the thousands of acres which it has already left behind it, and invading more and more the space occupied by the water. Innumerable plates of mica spangled the fine sand which the river brought down, and these, mixing with the water, and flashing like minute mirrors as the sun's rays fell upon them, gave the otherwise muddy stream a silvery appearance. Had I an opportunity I would make the following experiments:—

(*a*) Compare the colour of the light transmitted by a column of the lake water fifteen feet long with that transmitted by a second column, of the same length, derived from the melting of freshly fallen mountain snow.

(*b*) Compare in the same manner the colour of the ordinary water of the lake with that of the same water after careful distillation.

(*c*) Strictly examine whether the light transmitted by the ordinary water contains an excess of red over that transmitted by the distilled water: this latter point, as will be seen farther on, is one of peculiar interest.

The length is fixed at fifteen feet, because I have

The Lake of Geneva 33

found this length extremely efficient in similar experiments.

On returning to the pier at Villeneuve, a peculiar flickering motion was manifest upon the surface of the distant portions of the lake, and I soon noticed that the coast line was inverted by atmospheric refraction. It required a long distance to produce the effect: no trace of it was seen about the Castle of Chillon, but at Vevey and beyond it, the whole coast was clearly inverted; and

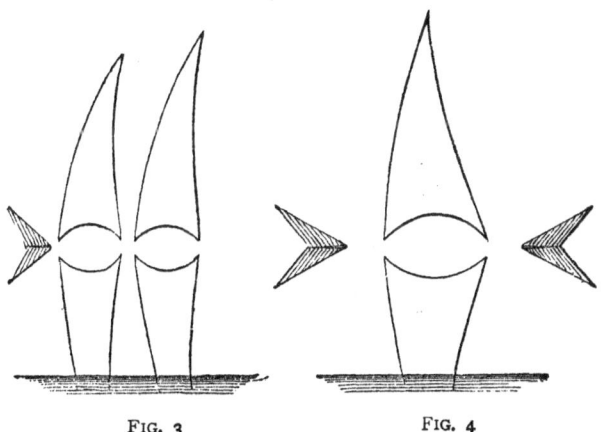

FIG. 3 FIG. 4

the houses on the margin of the lake were also imaged to a certain height. Two boats at a considerable distance presented the appearance sketched in Figs. 3 and 4; the hull of each, except a small portion at the end, was invisible, but the sails seemed lifted up high in the air, with their inverted images below; as the boats drew nearer the hulls appeared inverted, the apparent height of the vessel above the surface of the lake being thereby nearly doubled, while the sails and higher objects, in these cases, were almost completely cut away. When viewed through a telescope the sensible horizon of the lake presented a billowy tumultuous appearance, fragments being incessantly detached from it and suspended in the air.

The explanation of this effect is the same as that of the mirage of the desert, which may be found in almost any book on physics, and which so tantalised the French soldiers in Egypt. They often mistook this aërial inversion for the reflection from a lake, and on trial found hot and sterile sand at the place where they expected refreshing waters. The effect was shown by Monge, one of the learned men who accompanied the expedition, to be due to the total reflection of very oblique rays at the upper surface of the layer of rarefied air which was nearest to the heated earth. A sandy plain, in the early part of the day, is peculiarly favourable for the production of such effects; and on the extensive flat strand which stretches between Mont St. Michel and the coast adjacent to Avranches in Normandy, I have noticed Mont Tombeline reflected as if glass instead of sand surrounded it and formed its mirror.

CHAMOUNI AND THE MONTANVERT

V

On the evening of the 12th of July I reached Chamouni; the weather was not quite clear, but it was promising; white cumuli had floated round Mont Blanc during the day, but these diminished more and more, and the light of the setting sun was of that lingering rosy hue which bodes good weather. Two parallel beams of a purple tinge were drawn by the shadows of the adjacent peaks, straight across the Glacier des Bossons, and the Glacier des Pélerins was also steeped for a time in the same purple light. Once when the surrounding red illumination was strong, the shadows of the Grands Mulets falling upon the adjacent snow appeared of a *vivid green.*

This green belonged to the class of *subjective* colours, or colours produced by contrast, about which a volume might be written. The eye received the impression of

Chamouni and the Montanvert

green, but the colour was not external to the eye. Place a red wafer on white paper, and look at it intently, it will be surrounded in a little time by a green fringe: move the wafer bodily away, and the entire space which it occupied upon the paper will appear green. A body may have its proper colour entirely masked in this way. Let a red wafer be attached to a piece of red glass, and from a moderately illuminated position let the sky be regarded through the glass; the wafer will appear of a vivid green. If a strong beam of light be sent through a red glass and caused to fall upon a screen, which at the same time is moderately illuminated by a separate source of white light, an opaque body placed in the path of the beam will cast a green shadow upon the screen which may be seen by several hundred persons at once. If a blue glass be used, the shadow will be yellow, which is the complementary colour to blue.

When we suddenly pass from open sunlight to a moderately illuminated room, it appears dark at first, but after a little time the eye regains the power of seeing objects distinctly. Thus one effect of light upon the eye is to render it less sensitive, and light of any particular colour falling upon the eye blunts its appreciation *of that colour*. Let us apply this to the shadow upon the screen. This shadow is moderately illuminated by a jet of white light; but the space surrounding it is red, the effect of which upon the eye is to *blind* it in some degree to the perception of red. Hence, when the feeble white light of the shadow reaches the eye, the red component of this light is, as it were, abstracted from it, and the eye sees the residual colour, which is green. A similar explanation applies to the shadows of the Grands Mulets.

On the 13th of July I was joined by my friend Mr. Thomas Hirst, and on the 14th we examined together the end of the Mer de Glace. In former times the whole volume of the Arveiron escaped from beneath the ice at the end of the glacier, forming a fine arch at its place of issue. This year a fraction only of the water thus found egress; the greater portion of it escaping laterally from the glacier at the summit of the rocks called *Les Motets*, down which it tumbled in a fine cascade. The vault at the end of the glacier was nevertheless respectable, and rather tempting to a traveller in search of information regarding the structure of the ice. Perhaps, however, Nature meant to give me a friendly warning at the outset, for, while speculating as to the wisdom of entering the cavern, it suddenly gave way, and, with a crash which rivalled thunder, the roof strewed itself in ruins upon the floor.

Many years ago I had read with delight Coleridge's poem entitled 'Sunrise in the Valley of Chamouni,' and to witness in all perfection the scene described by the poet, I waited at Chamouni a day longer than was otherwise necessary. On the morning of Wednesday, the 15th of July, I rose before the sun; Mont Blanc and his wondrous staff of Aiguilles were without a cloud; eastward the sky was of a pale orange which gradually shaded off to a kind of rosy violet, and this again blended by imperceptible degrees with the deep zenithal blue. The morning star was still shining to the right, and the moon also turned a pale face towards the rising day. The valley was full of music; from the adjacent woods issued a gush of song, while the sound of the Arve formed a suitable bass to the shriller melody of the birds. The mountain rose for a time cold and grand, with no apparent stain upon his snows. Suddenly the sunbeams struck his crown and converted it into a boss of gold. For some time it remained the only gilded summit in view, holding communion with the dawn while all the others waited in silence. These, in the order of their heights, came afterwards, relaxing, as the sunbeams struck each in succession, into a blush and smile.

On the same day we had our luggage transported to the Montanvert, while we clambered along the lateral moraine of the glacier to the Chapeau. The rocks alongside the glacier were beautifully scratched and polished, and I paid particular attention to them, for the purpose of furnishing myself with a key to ancient glacier action. The scene to my right was one of the most wonderful I had ever witnessed. Along the entire slope of the

Glacier de Bois, the ice was cleft and riven into the most striking and fantastic forms. It had not yet suffered much from the wasting influence of the summer weather, but its towers and minarets sprang from the general mass with clean chiselled outlines. Some stood erect, others leaned, while the white débris, strewn here and there over the glacier, showed where the wintry edifices had fallen, breaking themselves to pieces, and grinding the masses on which they fell to powder. Some of them gave way during our inspection of the place, and shook the valley with the reverberated noise of their fall. I endeavoured to get near them, but failed; the chasms at the margin of the glacier were too dangerous, and the stones resting upon the heights too loosely poised to render persistence in the attempt excusable.

We subsequently crossed the glacier to the Montanvert, and I formally took up my position there. The rooms of the hotel were separated from each other by wooden partitions merely, and thus the noise of early risers in one room was plainly heard in the next. For the sake of quiet, therefore, I had my bed placed in the château next door,—a little octagonal building erected by some kind and sentimental Frenchman, and dedicated "*à la Nature.*" My host at first demurred, thinking the place not "*propre*," but I insisted, and he acquiesced. True the stone floor was dark with moisture, and on the walls a glistening was here and there observable, which suggested rheumatism, and other penalties, but I had had no experience of rheumatism, and trusted to the strength which mountain air and exercise were sure to give me, for power to resist its attacks. Moreover, to dispel some of the humidity, it was agreed that a large pine fire should be made there on necessary occasions.

Though singularly favoured on the whole, still our residence at the Montanvert was sufficiently long to give us specimens of all kinds of weather; and thus my château derived an interest from the mutations of external nature. Sometimes no breath disturbed the perfect serenity of the night, and the moon, set in a black-blue sky, turned a face of almost supernatural brightness to the mountains, while in her absence the thick-strewn stars alone flashed and twinkled through the transparent air. Sometimes dull dank fog choked the valley, and heavy rain plashed upon the stones outside. On two or three occasions we were favoured by a thunderstorm, every peal of which broke into a hundred echoes, while the seams of lightning which ran through the heavens produced a wonderful intermittence of gloom and glare. And as I sat within, musing on the experiences of the day, with my pine logs crackling, and the ruddy firelight gleaming over the walls, and lending animation to the visages sketched upon them with charcoal by the guides, I felt that my position was in every way worthy of a student of nature.

THE MER DE GLACE

VI

The name "Mer de Glace" has doubtless led many who have never seen this glacier to a totally erroneous conception of its character. Misled probably by this term, a distinguished writer, for example, defines a glacier to be a sheet of ice spread out upon the slope of a mountain; whereas the Mer de Glace is indeed a *river*, and not a *sea* of ice. But certain forms upon its surface, often noticed and described, and which I saw for the first time from the window of our hotel on the morning of the 16th of July, suggest at once the origin of the name. The glacier here has the appearance of a sea which, after it had been tossed by a storm, had suddenly stiffened into rest. The ridges upon its surface accurately resemble waves in shape, and this singular appearance is produced in the following way:—

Some distance above the Montanvert—opposite to the Echelets—the glacier, in passing down an incline, is rent by deep fissures, between each two of which a ridge of ice intervenes. At first the edges of these ridges are sharp and angular, but they are soon sculptured off by the action

The Mer de Glace 39

of the sun. The bearing of the Mer de Glace being approximately north and south, the sun at mid-day shines down the glacier, or rather very obliquely across it; and the consequence is, that the fronts of the ridges, which look downward, remain in shadow all the day, while the backs of the ridges, which look up the glacier, meet the direct stroke of the solar rays. The ridges thus acted upon have their hindmost angles wasted off and converted into slopes which represent the *back* of a wave, while the

FIG. 5

opposite sides of the ridges, which are protected from the sun, preserve their steepness, and represent the *front* of the wave. Fig. 5 will render my meaning at once plain.

The dotted lines are intended to represent three of the ridges into which the glacier is divided, with their interposed fissures; the dots representing the boundaries of the ridges when the glacier is first broken. The parallel shading lines represent the direction of the sun's rays, which, falling obliquely upon the ridges, waste away the right-hand corners, and finally produce wave-like forms.

We spent a day or two in making the general acquaintance of the glacier. On the 16th we ascended till we came to the rim of the Talèfre basin, from which we had a good view of the glacier system of the region. The laminated structure of the ice was a point which particularly interested me; and as I saw the exposed sections of the *névé*, counted the lines of stratification, and compared these with the lines upon the ends of the secondary glaciers, I felt the absolute necessity either of connecting the veined *structure* with the *strata* by a continuous chain of observations, or of proving by ocular evidence that they were totally distinct from each other. I was well acquainted with the literature of the subject, but nothing that I had read was sufficient to prove what I required. Strictly speaking, nothing that had been written upon the subject rose above the domain of *opinion*, while I felt that without absolute *demonstration* the question would never be set at rest.

Fig. 6

On this day we saw some fine glacier tables; flat masses of rock, raised high upon columns of ice: Fig. 6 is a sketch of one of the finest of them. Some of them fell from their pedestals while we were near them, and the clean ice-surfaces which they left behind sparkled with minute stars as the small bubbles of air ruptured the film of water by which they were overspread. I also noticed that "petit bruit de crépitation," to which M. Agassiz alludes, and which he refers to the rupture of the ice by the expansion of the air-bubbles contained within it. When I first read Agassiz's account of it, I thought it might be produced by the rupture of the minute air-bubbles which incessantly escape from the glacier. This, doubtless, produces an effect, but there is something in the character of the sound to be referred, I think, to a less obvious cause, which I shall notice further on.

At 6 P.M. this day I reached the Montanvert; and the same evening, wrapping my plaid around me, I wandered up towards Charmoz, and from its heights observed, as they had been observed fifteen years previously by Professor Forbes, the *dirt-bands* of the Mer de Glace. They were different from any I had previously seen, and I felt a strong desire to trace them to their origin. Content, however, with the performance of the day, and feeling healthily tired by it, I lay down upon the bilberry bushes and fell asleep. It was dark when I awoke, and I experienced some difficulty and risk in getting down from the petty eminence referred to.

The illumination of the glacier, as remarked by Professor Forbes, has great influcnce upon the appearance of the bands; they are best seen in a subdued light, and I think for the following reasons:—

The dirt-bands are seen simply because they send less light to the eye than the cleaner portions of the glacier which lie between them; two surfaces, differently illuminated, are presented to the eye, and it is found that this difference is more observable when the light is that of evening than when it is that of noon.

It is only within certain limits that the eye is able to perceive differences of intensity in different lights; beyond a certain intensity, if I may use the expression, light ceases to be light, and becomes mere pain. The naked eye can detect no difference in brightness between the electric light and the lime light, although, when we come to strict measurement, the former may possess many times the intensity of the latter. It follows from this that we might reduce the ordinary electric light to a fraction of its intensity, without any perceptible change of brightness to the naked eye which looks at it. But if we reduce the lime light in the same proportion the effect would be very different. This light lies much nearer to the limit at which the eye can appreciate differences of brightness, and its reduction might bring it quite within this limit, and make it sensibly dimmer than before. Hence we see that when two sources of intense light are presented to the eye, by reducing both the lights in the same proportion, the *difference* between them may become more perceptible.

Now the dirt-bands and the spaces between them resemble, in some measure, the two lights above mentioned. By the full glare of noon both are so strongly illuminated that the difference which the eye perceives is very small; as the evening advances the light of both is lowered in the same proportion, but the *differential* effect upon the eye is thereby augmented, and the bands are consequently more clearly seen.

VII

On Friday, the 17th of July, we commenced our measurements. Through the kindness of Sir Roderick Murchison, I found myself in the possession of an excellent five-inch theodolite, an instrument with the use of which both my friend Hirst and myself were perfectly familiar. We worked in concert for a few days to familiarise our assistant with the mode of proceeding, but afterwards it was my custom to simply determine the position where a measurement was to be made, and to leave the execution of it entirely to Mr. Hirst and our guide.

On the 20th of July I made a long excursion up the glacier, examining the moraines, the crevasses, the structure, the moulins, and the disintegration of the surface. I was accompanied by a boy named Edouard Balmat,[1] and found him so good an iceman that I was induced to take him with me on the following day also.

Looking upwards from the Montanvert to the left of the Aiguille de Charmoz, a singular gap is observed in the rocky mountain wall, in the centre of which stands a detached column of granite. Both cleft and pillar are shown in the frontispiece, to the right. The eminence to the left of this gap is signalised by Professor Forbes as

[1] "Le petit Balmat" my host always called him.

The Mer de Glace

one of the best stations from which to view the Mer de Glace, and this point, which I shall refer to hereafter as the *Cleft Station*, it was now my desire to attain. From the Montanvert side a steep gully leads to the cleft; up this couloir we proposed to try the ascent. At a considerable height above the Mer de Glace, and closely hugging the base of the Aiguille de Charmoz, is the small Glacier de Tendue, from which a steep slope stretches down to the Mer de Glace. This Tendue is the most *talkative* glacier I have ever known; the clatter of the small stones which fall from it is incessant. Huge masses of granite also frequently fall upon the glacier from the cliffs above it, and, being slowly borne downwards by the moving ice, are at length seen toppling above the terminal face of the glacier. The ice which supports them being gradually melted, they are at length undermined, and sent bounding down the slope with peal and rattle, according as the masses among which they move are large or small. The space beneath the glacier is cumbered with blocks thus sent down; some of them of enormous size.

The danger arising from this intermittent cannonade, though in reality small, has caused the guides to swerve from the path which formerly led across the slope to the promontory of Trélaporte. I say "small," because, even should a rock choose the precise moment at which a traveller is passing to leap down, the boulders at hand are so large and so capable of bearing a shock that the least presence of mind would be sufficient to place him in safety. But presence of mind is not to be calculated on under such circumstances, and hence the guides were right to abandon the path.

Reaching the mouth of our gully after a rough ascent, we took to the snow, instead of climbing the adjacent rocks. It was moist and soft, in fact in a condition altogether favourable for the "regelation" of its granules. As the foot pressed upon it the particles became cemented together. A portion of the pressure was transmitted laterally, which produced attachments beyond the boundary of the foot; thus as the latter sank, it pressed upon a surface which became continually wider and more rigid, and at length sufficiently strong to bear the entire weight of the body; the pressed snow formed in fact a virtual *camel's foot*, which soon placed a limit to the sinking. It is this same principle of regelation which enables men to cross snow bridges in safety. By gentle cautious pressure the loose granules of the substance are cemented into a continuous mass, all sudden shocks which might cause the frozen surfaces to snap asunder being avoided. In this way an arch of snow fifteen or twenty inches in thickness may be rendered so firm that a man will cross it, although it may span a chasm one hundred feet in depth.

As we ascended, the incline became very steep, and once or twice we diverged from the snow to the adjacent rocks; these were disintegrated, and the slightest disturbance was sufficient to bring them down; some fell, and from one of them I found it a little difficult to escape; for it grazed my leg, inflicting a slight wound as it passed. Just before reaching the cleft at which we aimed, the snow for a short distance was exceedingly steep, but we surmounted it; and I sat for a time beside the granite pillar, pleased to find that I could permit my legs to dangle over a precipice without prejudice to my head.

While we remained here a chamois made its appearance upon the rocks above us. Deeming itself too near, it climbed higher, and then turned round to watch us. It was soon joined by a second, and both formed a very pretty picture: their attitudes frequently changed, but they were always graceful; with head erect and horns curved back, a light limb thrown forward upon a ledge of rock, looking towards us with wild and earnest gaze, each seemed a type of freedom and agility. Turning now to the left, we attacked the granite tower, from which we purposed to scan the glacier, and were soon upon its top. My companion was greatly pleased —he was "très-content" to have reached the place—he felt assured that many old guides would have retreated from that ugly gully, with its shifting shingle and débris, and his elation reached its climax in the declaration that, if I resolved to ascend Mont Blanc without a guide, he was willing to accompany me.

From the position which we had attained, the prospect was exceedingly fine, both of the glaciers and of the mountains. Beside us was the Aiguille de Charmoz, piercing with its spikes of granite the clear air. To my mind it is one of the finest of the Aiguilles, noble in mass, with its summits singularly cleft and splintered. In some atmospheric colourings it has the exact appearance of a mountain of cast copper, and the manner in which some of its highest pinnacles are bent, suggesting the idea of ductility, gives strength to the illusion that the mass is metallic. At the opposite side of the glacier was the Aiguille Verte, with a cloud poised upon its point: it has long been the ambition of climbers to scale this peak, and on this day it was attempted by a young French count with a long retinue of guides He had not fair play, for before we quitted our position we heard the rumble of thunder upon the mountain, which indicated the presence of a foe more terrible than the avalanches themselves. Higher to the right, and also at the opposite side of the glacier, rose the Aiguille du Moine; and beyond was the basin of the Talèfre, the ice cascade issuing from which appeared, from our position, like the foam of a waterfall. Then came the Aiguille de Léchaud, the Petite Jorasse, the Grande Jorasse, and the Mont Tacul; all of which form a cradle for the Glacier de Léchaud. Mont Mallet, the Périades, and the Aiguille Noire, came next, and then the singular obelisk of the Aiguille du Géant, from which a serrated edge of cliff descends to the summit of the "Col."

Over the slopes of the Col du Géant was spread a coverlet of shining snow, at some places apparently as smooth as polished marble, at others broken so as to form precipices, on the pale blue faces of which the horizontal lines of bedding were beautifully drawn. As the eye approaches the line which stretches from the Rognon to the Aiguille Noire, the repose of the *névé* becomes more and more disturbed. Vast chasms are formed, which however are still merely indicative of the trouble in advance. If the glacier were lifted off we should probably see that the line just referred to would lie along the summit of a steep gorge; over this summit the glacier is pushed, and has its back periodically broken, thus forming vast transverse ridges which follow each other in succession down the slope. At the summit these ridges are often cleft by fissures transverse to *them*, thus forming detached towers of ice of the most picturesque and imposing character.[1] These towers often fall; and while some are caught upon the platforms of the cascade, others struggle with the slow energy of a behemoth through the débris which opposes them, reach the edges of the precipices which rise in succession along the fall, leap over, and, amid ice-smoke and thunder-peals, fight their way downwards.

A great number of secondary glaciers were in sight hanging on the steep slopes of the mountains, and from them streams sped downwards, falling over the rocks, and filling the valley with a low rich music. In front of me, for example, was the Glacier du Moine, and I could not help feeling as I looked at it, that the arguments drawn from the deportment of such glaciers against the "sliding theory," and which are still repeated in works upon the Alps, militate just as strongly against the "viscous theory." "How," demands the antagonist of the sliding theory, "can a secondary glacier exist upon so steep a slope? why does it not slide down as an avalanche?" "But how," the person addressed may retort, "can a mass which you assume to be viscous exist under similar conditions? If it be viscous, what prevents it from rolling down?" The sliding theory assumes the lubrication of the bed of the glacier, but on this cold height the quantity melted is too small to lubricate the bed, and hence the slow motion of these glaciers. Thus a sliding-theory man might reason, and, if the external deportment of secondary glaciers were to decide the question, De Saussure might perhaps have the best of the argument.

---

[1] To such towers the name *Séracs* is applied. In the châlets of Savoy, after the richer curd has been precipitated by rennet, a stronger acid is used to throw down what remains; an inferior kind of cheese called *Sérac* is thus formed, the shape and colour of which have suggested the application of the term to the cubical masses of ice.

The Mer de Glace

And with regard to the current idea, originated by M. de Charpentier, and adopted by Professor Forbes, that if a glacier slides it must slide as an avalanche, it may be simply retorted that, in part, *it does so;* but if it be asserted that an *accelerated motion* is the necessary motion of an avalanche, the statement needs qualification. An avalanche on passing through a rough couloir soon attains an uniform velocity — its motion being accelerated only up to the point when the sum of the resistances acting upon it is equal to the force drawing it downwards. These resistances are furnished by the numberless asperities which the mass encounters, and which incessantly check its descent, and render an accumulation of motion impossible. The motion of a man walking down stairs may be on the whole uniform, but it is really made up of an aggregate of small motions, each of which is accelerated; and it is easy to conceive how a glacier moving over an uneven bed, when released from one opposing obstacle will be checked by another, and its motion thus rendered sensibly uniform.

From the Aiguille du Géant and Les Périades a glacier descended, which was separated by the promontory of La Noire from the glacier proceeding from the Col du Géant. A small moraine was formed between them, which is marked *a* upon the diagram, Fig. 7. The great mass of the glacier descending from the Col du Géant came next, and this was bounded on the side nearest to Trélaporte by a small moraine *b*, the origin of which I could not see, its upper portion being shut out by a mountain promontory. Between the moraine *b* and the actual side of the valley was another little glacier, derived from some of the lateral tributaries. It was, however, between the moraines *a* and *b* that the great mass of the Glacier du Géant really lay. At the promontory of the Tacul the lateral moraines of the Glacier des Périades and of the Glacier de Léchaud united to form the medial moraine *c* of the Mer de Glace. Carrying the eye across the Léchaud, we had the moraine *d* formed by the union of the lateral moraines of the Léchaud and Talèfre; further to the left was the moraine *e*, which came from the Jardin, and beyond it was the second

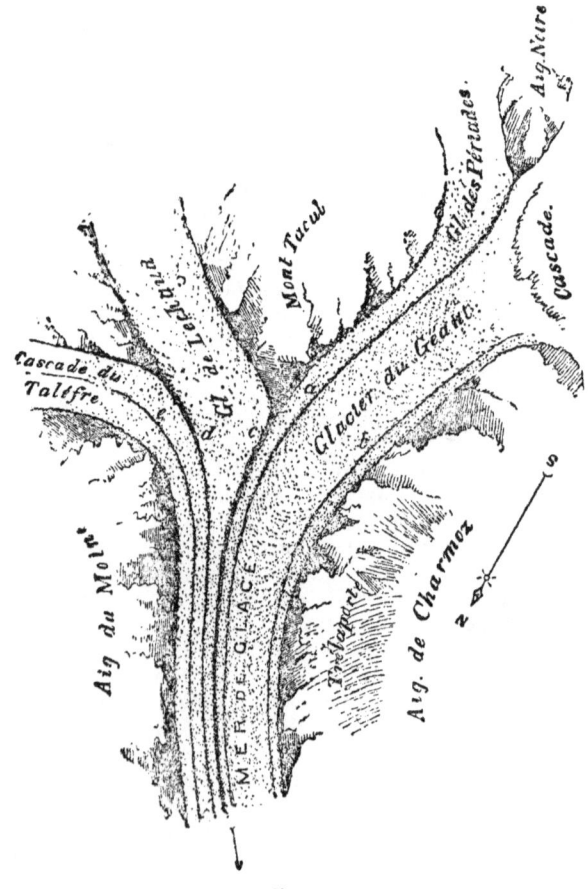

FIG. 7

lateral moraine of the Talèfre. The Mer de Glace is formed by the confluence of the whole of the glaciers

The Mer de Glace

here named; being forced at Trélaporte through a passage, the width of which appears considerably less than that of the single tributary, the Glacier du Géant.

In the ice near Trélaporte the blue veins of the glacier are beautifully shown; but they vary in distinctness according to the manner in which they are looked at. When regarded obliquely their colour is not so pronounced as when the vision plunges deeply into them. The weathered ice of the surface near Trélaporte could be cloven with great facility; I could with ease obtain plates of it a quarter of an inch thick, and possessing two square feet of surface. On the 28th of July I followed the veins several times from side to side across the Géant portion of the Mer de Glace; starting from one side, and walking along the veins, my route was directed obliquely downwards towards the axis of the tributary. At the axis I was forced to turn, in order to keep along the veins, and now ascended along a line which formed nearly the same angle with the axis at the other side. Thus the veins led me as it were along the two sides of a triangle, the vertex of which was near the centre of the glacier. The vertex was, however, in reality rounded off, and the figure rather resembled a hyperbola, which tended to coincidence with its asymptotes. This observation corroborates those of Professor Forbes with regard to the position of the veins, and, like him, I found that at the centre the veining, whose normal direction would be transverse to the glacier, was contorted and confused.

Near the side of the Glacier du Géant, above the promontory of Trélaporte, the ice is rent in a remarkable manner. Looking upwards from the lower portions of the glacier, a series of vertical walls, rising apparently one above the other, face the observer. I clambered up among these singular terraces, and now recognise, both from my sketch and memory, that their peculiar forms are due to the same action as that which has given their shape to the "billows" of the Mer de Glace. A series of profound crevasses is first formed. The Glacier du Géant deviates 14° from the meridian line, and hence the sun shines nearly down it during the middle portion of each day.

The backs of the ridges between the crevasses are thus rounded off, one boundary of each fissure is destroyed, or at least becomes a mere steep declivity, while the other boundary being shaded from the sun preserves its verticality; and thus a very curious series of precipices is formed.

Through all this dislocation, the little moraine on which I have placed the letter *b* in the sketch maintains its right to existence, and under it the laminated structure of this portion of the glacier appears to reach its most perfect development. The moraine was generally a mere dirt track, but one or two immense blocks of granite were perched upon it. I examined the ice underneath one of these, being desirous of seeing whether the pressure resulting from its enormous weight would produce a veining, but the result was not satisfactory. Veins were certainly to be seen in directions different from the normal ones, but whether they were due to the bending of the latter, or were directly owing to the pressure of the block, I could not say. The sides of a stream which had cut a deep gorge in the clean ice of the Glacier du Géant afforded a fine opportunity of observing the structure. It was very remarkable—highly significant indeed in a theoretic point of view. Two long and remarkably deep blue veins traversed the bottom of the stream, and bending upwards at a place where the rivulet curved, drew themselves like a pair of parallel lines upon the clean white ice. But the general structure was of a totally different character; it did not consist of long bars, but approximated to the lenticular form, and was, moreover, of a washy paleness, which scarcely exceeded in depth of colouring the whitish ice around.

To the investigator of the structure nothing can be finer than the appearance of the glacier from one of the ice terraces cut in the Glacier du Géant by its passage round Trélaporte. As far as the vision extended the dirt upon the surface of the ice was arranged in striæ. These striæ were not always straight lines, nor were they unbroken curves. Within slight limits the various parts into which a glacier is cut up by its crevasses enjoy a kind of independent motion. The grooves, for example, on two ridges which have been separated by a small fissure, may one day have their striæ perfect continuations of each other, but in a short time this identity of direction may be destroyed by a difference of motion between the ridges. Thus it is that the grooves upon the surface above Trélaporte are bent hither and thither, a crack or seam always marking the point where their continuity is ruptured. This bending occurs, however, within limits sufficiently small to enable the striæ to preserve the same general direction.

My attention had often been attracted this day by projecting masses of what at first appeared to be pure white snow, rising in seams above the general surface of the glacier. On examination, however, I found them to be compact ice, filled with innumerable air-cells, and so resistant as to maintain itself in some places at a height of four feet above the general level. When amongst the ridges they appeared discontinuous and confused, being scattered apparently at random over the glacier; but when viewed from a sufficient distance, the detached parts showed themselves to belong to a system of white seams which swept quite across the Glacier du Géant, in a direction concentric with the structure. Unable to account for these singular seams, I climbed up among the tributary glaciers on the Rognon side of the Glacier du Géant, and remained there until the sun sank behind the neighbouring peaks, and the fading light warned me that it was time to return.

VIII

Early on the following day I was again upon the ice. I first confined myself to the right side of the Glacier du Géant, and found that the veins of white ice which I had noticed on the previous day were exclusively confined to this glacier, or to the space between the moraines a and b (Fig. 7), bending up so that the moraine a between the Glacier du Géant and the Glacier des Périades was tangent to them. At a good distance up the glacier I encountered a considerable stream rushing across it almost from side to side. I followed the rivulet, examining the sections which it exposed. At a certain point three other streams united, and formed at their place of confluence a small green lake. From this a rivulet rushed, which was joined by the stream whose track I had pursued, and at this place of junction a second green lake was formed, from which flowed a stream equal in volume to the sum of all the tributaries. It entered a crevasse, and took the bottom of the fissure for its bed. Standing at the entrance of the chasm, a low muffled thunder resounding through the valley attracted my attention. I followed the crevasse, which deepened and narrowed, and, by the blue light of the ice, could see the stream gambolling along its bottom, and flashing as it jumped over the ledges which it encountered in its way. The fissure at length came to an end: placing a foot on each side of it, and withholding the stronger light from my eyes, I looked down between its shining walls, and saw the stream plunge into a shaft which carried it to the bottom of the glacier.

Slowly, and in zigzag fashion, as the crevasses demanded, I continued to ascend, sometimes climbing vast humps of ice from which good views of the surrounding glacier were obtained; sometimes hidden in the hollows between the humps, in which also green glacier tarns were often formed, very lonely and very beautiful.

While standing beside one of these, and watching the moving clouds which it faithfully mirrored, I heard the sound of what appeared to be a descending avalanche, but the time of its continuance surprised me. Looking through my opera-glass in the direction of the sound, I saw issuing from the end of a secondary glacier on the Tacul side a torrent of what appeared to me to be stones and mud. I could see the stones and finer débris jumping down the declivities, and shaping themselves into singular cascades. The noise continued for a quarter of an hour, after which the torrent rapidly diminished,

The Mer de Glace

until, at length, the ordinary little stream due to the melting of the glacier alone remained. A subglacial lake had burst its boundary, and carried along with it in its rush downwards the débris which it met with in its course.

In some places I found the crevasses difficult, the ice being split in a very singular manner. Vast plates of it not more than a foot in thickness were sometimes detached from the sides of the crevasses, and stood alone. I was now approaching the base of the *séracs*, and the glacier around me still retained a portion of the turbulence of the cascade. I halted at times amid the ruin and confusion, and examined with my glass the cascade itself. It was a wild and wonderful scene, suggesting throes of spasmodic energy, though, in reality, all its dislocation had been *slowly* and *gradually* produced. True, the stratified blocks which here and there cumbered the terraces suggested *débacles*, but these were local and partial, and did not affect the general question. There is scarcely a case of geological disturbance which could not be matched with its analogue upon the glaciers,—contortions, faults, fissures, joints, and dislocations,—but in the case of the ice we can prove the effects to be due to slowly acting causes; how reasonable is it then to ascribe to the operation of similar causes, which have had an incomparably longer time to work, many geological effects which at first sight might suggest sudden convulsion!

Wandering slowly upwards, successive points of attraction drawing me almost unconsciously on, I found myself as the day was declining deep in the entanglements of the ice. A shower commenced, and a splendid rainbow threw an oblique arch across the glacier. I was quite alone; the scene was exceedingly impressive, and the possibility of difficulties on which I had not calculated intervening between me and the lower glacier, gave a tinge of anxiety to my position. I turned towards home; crossed some bosses of ice and rounded others; I followed the tracks of streams which were very irregular on this portion of the glacier, bending hither and thither, rushing through deep-cut channels, falling in cascades and expanding here and there to deep green lakes; they often plunged into the depths of the ice, flowed under it with hollow gurgle, and reappeared at some distant point. I threaded my way cautiously amid systems of crevasses, scattering with my axe, to secure a footing, the rotten ice of the sharper crests, which fell with a ringing sound into the chasms at either side. Strange subglacial noises were sometimes heard, as if caverns existed underneath, into which blocks of ice fell at intervals, transmitting the shock of their fall with a dull boom to the surface of the glacier. By the steady surmounting of difficulties one after another, I at length placed them all behind me, and afterwards hastened swiftly along the glacier to my mountain home.

On the 30th incessant rain confined us to indoor work; on the 31st we determined the velocity with which the glacier is forced through the entrance of the trunk valley at Trélaporte, and also the motion of the Grand Moulin. We also determined both the velocity and the width of the Glacier du Géant. The 1st of August was spent by me at the cascade of the Talèfre, examining the structure, crumpling, and scaling off of the ice. Finding that the rules at Chamouni put an unpleasant limit to my demands on my guide Simond, I visited the Guide Chef on the 2nd of August, and explained to him the object of my expedition, pointing out the inconvenience which a rigid application of the rules made for tourists would impose upon me. He had then the good sense to acknowledge the reasonableness of my remarks, and to grant me the liberty I requested. The 3rd of August was employed in determining the velocity and width of the Glacier de Léchaud, and in observations on the lamination of the glacier.

THE JARDIN

IX

On the 4th of August, with a view of commencing a series of observations on the inclinations of the Mer de Glace and its tributaries, we had our theodolite transported to the *Jardin*, which, as is well known, lies like an island in the middle of the Glacier du Talèfre. We reached the place by the usual route, and found some tourists reposing on the soft green sward which covers the lower portion, and to which, and the flowers which spangle it, the place owes its name. Towards the summit of the Jardin, a rock jutted forward, apparently the very apex of the place, or at least hiding by its prominence everything that might exist behind it; leaving our guide with the instrument, we aimed at this, and soon left the grass and flowers behind us. Stepping amid broken fragments of rock, along slopes of granite, with fat felspar crystals which gave the boots a hold, and crossing at intervals patches of snow, which continued still to challenge the summer heat, I at length found myself upon the peak referred to; and, although it was not the highest, the unimpeded view which it commanded induced me to get astride it. The Jardin was completely encircled by the ice of the glacier, and this was held in a mountain basin, which was bounded all round by a grand and cliffy rim. The outline of the dark brown crags—a deeply serrated and irregular line—was forcibly drawn against the blue heaven, and still more strongly against some white and fleecy clouds which lay here and there behind it; while detached spears and pillars of rock, sculptured by frost and lightning, stood like a kind of defaced statuary along the ridge. All round the basin the snow reared itself like a buttress against the precipitous cliffs, being streaked and fluted by the descent of blocks from the summits. This mighty tub is the collector of one of the tributaries of the Mer de Glace. According to Professor Forbes, its greatest diameter is 4200 yards, and out of it the half-formed ice is squeezed through a precipitous gorge about 700 yards wide, forming there the ice cascade of the Talèfre. Bounded on one side by the Grande Jorasse, and on the other by Mont Mallet, the principal tributary of the Glacier de Léchaud lay white and pure upon the mountain slope. Round further to the right we had the vast plateau whence the Glacier du Géant is fed, fenced on the left by the Aiguille du Géant and the Aiguille Noire, and on the right by the Monts Maudits and Mont Blanc. The scene was a truly majestic one. The mighty Aiguilles piercing the sea of air, the soft white clouds floating here and there behind them; the shining snow with its striped faults and precipices; the deep blue firmament overhead; the peals of avalanches and the sound of water;—all conspired to render the scene glorious, and our enjoyment of it deep.

A voice from above hailed me as I moved from my perch; it was my friend, who had found a lodgment upon the edge of a rock which was quite detached from the Jardin, being the first to lift its head in opposition to the descending *névé*. Making a détour round a steep concave slope of the glacier, I reached the flat summit of the rock. The end of a ridge of ice abutted against it, which was split and bent by the pressure so as to form a kind of arch. I cut steps in the ice, and ascended until I got beneath the azure roof. Innumerable little rills of pellucid water descended from it. Some came straight down, clear for a time, and apparently motionless, rapidly tapering at first, and more slowly afterwards, until, at the point of maximum contraction, they resolved themselves into strings of liquid pearls which pattered against the ice floor underneath. Others again, owing to the directions of the little streamlets of which they were constituted, formed spiral figures of great beauty: one liquid vein wound itself round another, forming a spiral protuberance, and owing to the centrifugal motion thus imparted, the vein, at its place of rupture, scattered itself laterally in little liquid spherules.[1] Even at this great elevation the structure of the ice was fairly developed, not with the sharpness to be observed lower down, but still perfectly decided. Blue bands crossed the ridge of ice to which I have referred, at right angles to the direction of the pressure.

I descended, and found my friend beneath an overhanging rock. Immediately afterwards a peal like that of thunder shook the air, and right in front of us an avalanche darted down the brown cliffs, then along a steep

---

[1] The recent hydraulic researches of Professor Magnus furnish some beautiful illustrations of this action.

slope of snow which reared itself against the mountain wall, carrying with it the débris of the rocks over which it passed, until it finally lay a mass of sullied rubbish at the base of the incline: the whole surface of the Talèfre is thus soiled. Another peal was heard immediately afterwards, but the avalanche which caused it was hidden from us by a rocky promontory. From this same promontory the greater portion of the medial moraine which descends the cascade of the Talèfre is derived, forming at first a gracefully winding curve, and afterwards stretching straight to the summit of the fall. In the chasms of the cascade its boulders are engulfed, but the lost moraine is restored below the fall, as if disgorged by the ice which had swallowed it. From the extremity of the Jardin itself a mere driblet of a moraine proceeds, running parallel to the former, and like it disappearing at the summit of the cascade.

We afterwards descended towards the cascade, but long before this is attained the most experienced iceman would find himself in difficulty. Transverse crevasses are formed, which follow each other so speedily as to leave between them mere narrow ridges of ice, along which we moved cautiously, jumping the adjacent fissures, or getting round them, as the case demanded. As we approached the jaws of the gorge, the ridges dwindled to mere plates and wedges, which being bent and broken by the lateral pressure, added to the confusion, and warned us not to advance. The position was in some measure an exciting one. Our guide had never been here before; we were far from the beaten track, and the riven glacier wore an aspect of treacherous hostility. As at the base of the *séracs*, a subterranean noise sometimes announced the falling of ice-blocks into hollows underneath, the existence of which the resonant concussion of the fallen mass alone revealed. There was thus a dash of awe mingled with our thoughts; a stirring up of the feelings which troubled the coolness of the intellect. We finally swerved to the right, and by a process the reverse of straightforward reached the Couvercle. Nightfall found us at the threshold of our hotel.

X

. On the 5th we were engaged for some time in an important measurement at the Tacul. We afterwards ascended towards the *séracs*, and determined the inclinations of the Glacier du Géant downwards. Dense cloud-masses gathered round the points of the Aiguilles, and the thunder bellowed at intervals from the summit of Mont Blanc. As we descended the Mer de Glace the valley in front of us was filled with a cloud of pitchy darkness. Suddenly from side to side this field of gloom was riven by a bar of lightning of intolerable splendour; it was followed by a peal of commensurate grandeur, the echoes of which leaped from cliff to cliff long after the first sound had died away. The discharge seemed to unlock the clouds above us, for they showered their liquid spheres down upon us with a momentum like that of swan-shot: all the way home we were battered by this pellet-like rain. On the 6th the rain continued with scarcely any pause; on the 7th I was engaged all day upon the Glacier du Géant; on the morning of the 8th heavy hail had fallen there, the stones being perfect spheres; the rounded rain-drops had solidified during their descent without sensible change of form. When this hail was squeezed together, it exactly resembled a mass of oolitic limestone which I had picked up in 1853 near Blankenburg in the Harz. Mr. Hirst and myself were engaged together this day taking the inclinations: he struck his theodolite at the Angle, and went home accompanied by Simond, and, the evening being extremely serene, I pursued my way down the centre of the glacier towards the Echelets. The crevasses as I advanced became more deep and frequent, the ridges of ice between them becoming gradually narrower. They were very fine, their downward faces being clear cut, perfectly vertical, and in many cases beautifully veined. Vast plates of ice moreover often stood out midway between the walls of the chasms, as if cloven from the glacier and afterwards set on edge. The place was certainly one calculated to test the skill and nerve of an iceman; and as the day drooped, and the shadow in the valley deepened, a feeling approaching to awe took possession of me. My route was an exaggerated zigzag; right and left amid the chasms wherever a hope of progress opened; and here I made the experience which I have often repeated since, and laid to heart as regards intellectual work also, that enormous difficulties may be overcome when they are attacked in earnest. Sometimes I found myself so hedged in by fissures that escape seemed absolutely impossible; but close and resolute examination so often revealed a means of exit, that I felt in all its force the brave verity of the remark of Mirabeau, that the word "impossible" is a mere blockhead of a word. It finally became necessary to reach the shore, but I found this a work of extreme difficulty. At length, however, it became pretty evident that, if I could cross a certain crevasse, my retreat would be secured. The width of the fissure seemed to be fairly within jumping distance, and if I could have calculated on a safe purchase for my foot I should have thought little of the spring; but the ice on the edge from which I was to leap was loose and insecure, and hence a kind of nervous thrill shot through me as I made the bound. The opposite side was fairly reached, but an involuntary tremor shook me all over after I felt myself secure. I reached the edge of the glacier without further serious difficulty, and soon after found myself steeped in the creature comforts of our hotel.

On Monday, 10th August, I had the great pleasure of being joined by my friend Huxley; and though the weather was very unpromising, we started together up the glacier, he being desirous to learn something of its general features, and, if possible, to reach the Jardin. We reached the Couvercle, and squeezed ourselves through the Egralets; but here the rain whizzed past us, and dense fog settled upon the cascade of the Talèfre, obscuring all its parts. We met Mr. Galton, the African traveller, returning from an attempt upon the Jardin; and learning that his guides had lost their way in the fog, we deemed it prudent to return.

The foregoing brief notes will have informed the reader that at the period of Mr. Huxley's arrival I was not without due training upon the ice; I may also remark, that on the 25th of July I reached the summit of the Col du Géant, accompanied by the boy Balmat, and returned to the Montanvert on the same day. My health was perfect, and incessant practice had taught me the art of dealing with the difficulties of the ice. From the time of my arrival at the Montanvert the thought of ascending Mont Blanc, and thus expanding my knowledge of the glaciers, had often occurred to me, and I think I was justified in feeling that the discipline which both my friend Hirst and myself had undergone ought to enable us to accomplish the journey in a much more modest way than ordinary. I thought a single guide sufficient for this purpose, and I was strengthened in this opinion by the fact that Simond, who was a man of the strictest prudence, and who at first declared four guides to be necessary, had lowered his demand first to two, and was now evidently willing to try the ascent with us alone.

On mentioning the thing to Mr. Huxley he at once resolved to accompany us. On the 11th of August the weather was exceedingly fine, though the snow which had fallen during the previous days lay thick upon the glacier. At noon we were all together at the Tacul, and the subject of attempting Mont Blanc was mooted and discussed. My opinion was that it would be better to wait until the fresh snow which loaded the mountain had disappeared; but the weather was so exquisite that my friends thought it best to take advantage of it. We accordingly entered into an agreement with our guide, and immediately descended to make preparations for commencing the expedition on the following morning.

FIRST ASCENT OF MONT BLANC, 1857

XI

On Wednesday, the 12th of August, we rose early, after a very brief rest on my part. Simond had proposed to go down to Chamouni, and commence the ascent in the

First Ascent of Mont Blanc, 1857

usual way, but we preferred crossing the mountains from the Montanvert, straight to the Glacier des Bossons. At eight o'clock we started, accompanied by two porters who were to carry our provisions to the Grands Mulets. Slowly and silently we climbed the hill-side towards Charmoz. We soon passed the limits of grass and rhododendrons, and reached the slabs of gneiss which overspread the summit of the ridge, lying one upon the other like coin upon the table of a money-changer. From the highest point I turned to have a last look at the Mer de Glace; and through a pair of very dark spectacles I could see with perfect distinctness the looped dirt-bands of the glacier, which to the naked eye are scarcely discernible except by twilight. Flanking our track to the left rose a series of mighty Aiguilles—the Aiguille de Charmoz, with its bent and rifted pinnacles; the Aiguille du Grepon, the Aiguille de Blaitière, the Aiguille du Midi, all piercing the heavens with their sharp pyramidal summits. Far in front of us rose the grand snow-cone of the Dôme du Gouté, while, through a forest of dark pines which gathered like a cloud at the foot of the mountain, gleamed the white minarets of the Glacier des Bossons. Below us lay the Valley of Chamouni, beyond which were the Brevent and the chain of the Aiguilles Rouges; behind us was the granite obelisk of the Aiguille du Dru, while close at hand science found a corporeal form in a pyramid of stones used as a trigonometrical station by Professor Forbes. Sound is known to travel better up hill than down, because the pulses transmitted from a denser medium to a rarer, suffer less loss of intensity than when the transmission is in the opposite direction; and now the mellow voice of the Arve came swinging upwards from the heavier air of the valley to the lighter air of the hills in rich deep cadences.

The way for a time was excessively rough, our route being overspread with the fragments of peaks which had once reared themselves to our left, but which frost and lightning had shaken to pieces, and poured in granite avalanches down the mountain. We were sometimes among huge angular boulders, and sometimes amid lighter shingle, which gave way at every step, thus forcing us to shift our footing incessantly. Escaping from these, we crossed the succession of secondary glaciers which lie at the feet of the Aiguilles, and having secured firewood found ourselves after some hours of hard work at the Pierre l'Echelle. Here we were furnished with leggings of coarse woollen cloth to keep out the snow; they were tied under the knees and quite tightly again over the insteps, so that the legs were effectually protected. We had some refreshment, possessed ourselves of the ladder, and entered upon the glacier.

The ice was excessively fissured: we crossed crevasses and crept round slippery ridges, cutting steps in the ice wherever climbing was necessary. This rendered our progress very slow. Once, with the intention of lending a helping hand, I stepped forward upon a block of granite which happened to be poised like a rocking stone upon the ice, though I did not know it; it treacherously turned under me; I fell, but my hands were in instant requisition, and I escaped with a bruise, from which, however, the blood oozed angrily. We found the ladder necessary in crossing some of the chasms, the iron spikes at its end being firmly driven into the ice at one side, while the other end rested on the opposite side of the fissure. The middle portion of the glacier was not difficult. Mounds of ice rose beside us right and left, which were sometimes split into high towers and gaunt-looking pyramids, while the space between was unbroken. Twenty minutes' walking brought us again to a fissured portion of the glacier, and here our porter left the ladder on the ice behind him. For some time I was not aware of this, but we were soon fronted by a chasm to pass which we were in consequence compelled to make a long and dangerous circuit amid crests of crumbling ice. This accomplished, we hoped that no repetition of the process would occur, but we speedily came to a second fissure, where it was necessary to step from a projecting end of ice to a mass of soft snow which overhung the opposite side. Simond could reach this snow with his long-handled axe; he beat it down to give it rigidity, but it was exceedingly tender, and as he worked at it he continued to express his fears that it would

First Ascent of Mont Blanc, 1857

not bear us. I was the lightest of the party, and therefore tested the passage first; being partially lifted by Simond on the end of his axe, I crossed the fissure, obtained some anchorage at the other side, and helped the others over. We afterwards ascended until another chasm, deeper and wider than any we had hitherto encountered, arrested us. We walked alongside of it in search of a snow bridge, which we at length found, but the keystone of the arch had unfortunately given way, leaving projecting eaves of snow at both sides, between which we could look into the gulf, till the gloom of its deeper portions cut the vision short. Both sides of the crevasse were sounded, but no sure footing was obtained; the snow was beaten and carefully trodden down as near to the edge as possible, but it finally broke away from the foot and fell into the chasm. One of our porters was short-legged and a bad iceman; the other was a daring fellow, and he now threw the knapsack from his shoulders, came to the edge of the crevasse, looked into it, but drew back again. After a pause he repeated the act, testing the snow with his feet and staff. I looked at the man as he stood beside the chasm manifestly undecided as to whether he should take the step upon which his life would hang, and thought it advisable to put a stop to such perilous play. I accordingly interposed, the man withdrew from the crevasse, and he and Simond descended to fetch the ladder.

While they were away Huxley sat down upon the ice, with an expression of fatigue stamped upon his countenance: the spirit and the muscles were evidently at war, and the resolute will mixed itself strangely with the sense of peril and feeling of exhaustion. He had been only two days with us, and, though his strength is great, he had had no opportunity of hardening himself by previous exercise upon the ice for the task which he had undertaken. The ladder now arrived, and we crossed the crevasse. I was intentionally the last of the party, Huxley being immediately in front of me. The determination of the man disguised his real condition from everybody but myself, but I saw that the exhausting journey over the boulders and débris had been too much for his London limbs. Converting my waterproof haversack into a cushion, I made him sit down upon it at intervals, and by thus breaking the steep ascent into short stages we reached the cabin of the Grands Mulets together. Here I spread a rug on the boards, and placing my bag for a pillow, he lay down, and after an hour's profound sleep he rose refreshed and well; but still he thought it wise not to attempt the ascent farther. Our porters left us: a baton was stretched across the room over the stove, and our wet socks and leggings were thrown across it to dry; our boots were placed around the fire, and we set about preparing our evening meal. A pan was placed upon the fire, and filled with snow, which in due time melted and boiled; I ground some chocolate and placed it in the pan, and afterwards ladled the beverage into the vessels we possessed, which consisted of two earthen dishes and the metal cases of our brandy flasks. After supper Simond went out to inspect the glacier, and was observed by Huxley, as twilight fell, in a state of deep contemplation beside a crevasse.

Gradually the stars appeared, but as yet no moon. Before lying down we went out to look at the firmament, and noticed, what I suppose has been observed to some extent by everybody, that the stars near the horizon twinkled busily, while those near the zenith shone with a steady light. One large star in particular excited our admiration; it flashed intensely, and changed colour incessantly, sometimes blushing like a ruby, and again gleaming like an emerald. A determinate colour would sometimes remain constant for a sensible time, but usually the flashes followed each other in very quick succession. Three planks were now placed across the room near the stove, and upon these, with their rugs folded round them, Huxley and Hirst stretched themselves, while I nestled on the boards at the most distant end of the room. We rose at eleven o'clock, renewed the fire and warmed ourselves, after which we lay down again. I at length observed a patch of pale light upon the wooden wall of the cabin, which had entered through a hole in the end of the edifice, and rising found that it was past one o'clock. The cloudless moon was shining over the wastes of snow, and the scene outside was at once wild, grand, and beautiful.

Breakfast was soon prepared, though not without difficulty; we had no candles, they had been forgotten; but I fortunately possessed a box of wax matches, of which Huxley took charge, patiently igniting them in succession, and thus giving us a tolerably continuous light. We had some tea, which had been made at the Montanvert, and carried to the Grands Mulets in a bottle. My memory of that tea is not pleasant; it had been left a whole night in contact with its leaves, and smacked strongly of tannin. The snow-water, moreover, with which we diluted it was not pure, but left a black residuum at the bottom of the dishes in which the beverage was served. The few provisions deemed necessary being placed in Simond's knapsack, at twenty minutes past two o'clock we scrambled down the rocks, leaving Huxley behind us.

The snow was hardened by the night's frost, and we were cheered by the hope of being able to accomplish the ascent with comparatively little labour. We were environed by an atmosphere of perfect purity; the larger stars hung like gems above us, and the moon, about half full, shone with wondrous radiance in the dark firmament. One star in particular, which lay eastward from the moon, suddenly made its appearance above one of the Aiguilles, and burned there with unspeakable splendour. We turned once towards the Mulets, and saw Huxley's form projected against the sky as he stood upon a pinnacle of rock; he gave us a last wave of the hand and descended, while we receded from him into the solitudes.

The evening previous our guide had examined the glacier for some distance, his progress having been arrested by a crevasse. Beside this we soon halted: it was spanned at one place by a bridge of snow, which was of too light a structure to permit of Simond's testing it alone; we therefore paused while our guide uncoiled a rope and tied us all together. The moment was to me a peculiarly solemn one. Our little party seemed so lonely and so small amid

E

the silence and the vastness of the surrounding scene. We were about to try our strength under unknown conditions, and as the various possibilities of the enterprise crowded on the imagination, a sense of responsibility for a moment oppressed me. But as I looked aloft and saw the glory of the heavens, my heart lightened, and I remarked cheerily to Hirst that Nature seemed to smile upon our work. "Yes," he replied, in a calm and earnest voice, "and, God willing, we shall accomplish it."

A pale light now overspread the eastern sky, which increased, as we ascended, to a daffodil tinge; this afterwards heightened to orange, deepening at one extremity into red, and fading at the other into a pure ethereal hue to which it would be difficult to assign a special name. Higher up the sky was violet, and this changed by insensible degrees into the darkling blue of the zenith, which had to thank the light of moon and stars alone for its existence. We wound steadily for a time through valleys of ice, climbed white and slippery slopes, crossed a number of crevasses, and after some time found ourselves beside a chasm of great depth and width, which extended right and left as far as we could see. We turned to the left, and marched along its edge in search of a *pont;* but matters became gradually worse: other crevasses joined on to the first one, and the further we proceeded the more riven and dislocated the ice became. At length we reached a place where further advance was impossible. Simond in his difficulty complained of the want of light, and wished us to wait for the advancing day; I, on the contrary, thought that we had light enough and ought to make use of it. Here the thought occurred to me that Simond, having been only once before to the top of the mountain, might not be quite clear about the route; the glacier, however, changes within certain limits from year to year, so that a general knowledge was all that could be expected, and we trusted to our own muscles to make good any mistake in the way of guidance. We now turned and retraced our steps along the edges of chasms where the ice was disintegrated and insecure, and succeeded at length in finding a bridge which bore us across the crevasse. This error caused us the loss of an hour, and after walking for this time we could cast a stone from the point we had attained to the place whence we had been compelled to return.

Our way now lay along the face of a steep incline of snow, which was cut by the fissure we had just passed, in a direction parallel to our route. On the heights to our right, loose ice-crags seemed to totter, and we passed two tracks over which the frozen blocks had rushed some short time previously. We were glad to get out of the range of these terrible projectiles, and still more so to escape the vicinity of that ugly crevasse. To be killed in the open air would be a luxury, compared with having the life squeezed out of one in the horrible gloom of these chasms. The blush of the coming day became more and more intense; still the sun himself did not appear, being hidden from us by the peaks of the Aiguille du Midi, which were drawn clear and sharp against the brightening sky. Right under this Aiguille were heaps of snow smoothly rounded and constituting a portion of the sources whence the Glacier du Géant is fed; these, as the day advanced, bloomed with a rosy light. We reached the Petit Plateau, which we found covered with the remains of ice avalanches; above us upon the crest of the mountain rose three mighty bastions, divided from each other by deep vertical rents, with clean smooth walls, across which the lines of annual bedding were drawn like courses of masonry. From these, which incessantly renew themselves, and from the loose and broken ice-crags near them, the boulders amid which we now threaded our way had been discharged. When they fall their descent must be sublime.

The snow had been gradually getting deeper, and the ascent more wearisome, but superadded to this at the Petit Plateau was the uncertainty of the footing between the blocks of ice. In many places the space was merely covered by a thin crust, which, when trod upon, instantly yielded, and we sank with a shock sometimes to the hips. Our way next lay up a steep incline to the Grand Plateau, the depth and tenderness of the snow augmenting as we ascended. We had not yet seen the sun, but, as we attained the brow which forms the entrance to the Grand Plateau, he hung his disk upon a spike of rock to our left, and, surrounded by a glory of interference spectra of the most gorgeous colours, blazed down upon us. On the Grand Plateau we halted and had our frugal refreshment. At some distance to our left was the crevasse into which Dr. Hamel's three guides were precipitated by an avalanche in 1820; they are still entombed in the ice, and some future explorer may perhaps see them disgorged lower down, fresh and undecayed. They can hardly reach the surface until they pass the snow-line of the glacier, for above this line the quantity of snow that annually falls being in excess of the quantity melted, the tendency would be to make the ice-covering above them thicker. But it is also possible that the waste of the ice underneath may have brought the bodies to the bed of the glacier, where their very bones may have been ground to mud by an agency which the hardest rocks cannot withstand.

As the sun poured his light upon the Plateau the little snow-facets sparkled brilliantly, sometimes with a pure white light, and at others with prismatic colours. Contrasted with the white spaces above and around us were the dark mountains on the opposite side of the valley of Chamouni, around which fantastic masses of cloud were beginning to build themselves. Mont Buet, with its cone of snow, looked small, and the Brevent altogether mean; the limestone bastions of the Fys, however, still presented a front of gloom and grandeur. We traversed the Grand Plateau, and at length reached the base of an extremely steep incline which stretched upwards towards the Corridor. Here, as if produced by a fault, consequent upon the sinking of the ice in front, rose a vertical precipice, from the coping of which vast stalactites of ice depended. Previous to reaching this place I had noticed a haggard expression upon the countenance of our guide, which was now intensified by the prospect of the ascent before him. Hitherto he had always been in front, which was certainly the most fatiguing position. I felt that I must now take the lead, so I spoke cheerily to the man and placed him behind me. Marking a number of points upon the slope as resting places, I went swiftly from one to the other. The surface of the snow had been partially melted by the sun and then refrozen, thus forming a superficial crust, which bore the weight up to a certain point, and then suddenly gave way, permitting the leg to sink to above the knee. The shock consequent on this, and the subsequent effort necessary to extricate the leg, were extremely fatiguing. My motion was complained of as too quick, and my tracks as imperfect; I moderated the former, and to render my footholes broad and sure, I stamped upon the frozen crust, and twisted my legs in the soft mass underneath,—a terribly exhausting process. I thus led the way to the base of the Rochers Rouges, up to which the fault already referred to had prolonged itself as a crevasse, which was roofed at one place by a most dangerous-looking snow-bridge. Simond came to the front; I drew his attention to the state of the snow, and proposed climbing the Rochers Rouges; but, with a promptness unusual with him, he replied that this was impossible; the bridge was our only means of passing, and we must try it. We grasped our ropes, and dug our feet firmly into the snow to check the man's descent if the *pont* gave way, but to our astonishment it bore him, and bore us safely after him. The slope which we had now to ascend had the snow swept from its surface, and was therefore firm ice. It was most dangerously steep, and, its termination being the fretted coping of the precipice to which I have referred, if we slid downwards we should shoot over this and be dashed to pieces upon the ice below.[1] Simond, who had come to the front to cross the crevasse, was now engaged in cutting steps, which he made deep and large, so that they might serve us on our return. But the listless strokes of his axe proclaimed his exhaustion; so I took the implement out of his hands, and changed places with him. Step after step was hewn, but the top of the Corridor appeared ever to recede from us. Hirst was behind

---

[1] Those acquainted with the mountain will at once recognise the grave error here committed. In fact, on starting from the Grands Mulets we had crossed the glacier too far, and throughout were much too close to the Dôme du Gouté.

unoccupied, and could thus turn his thoughts to the peril of our position: he *felt* the angle on which we hung, and saw the edge of the precipice, to which less than a quarter of a minute's slide would carry us, and for the first time during the journey he grew giddy. A cigar which he lighted for the purpose tranquillised him.

I hewed sixty steps upon this slope, and each step had cost a minute, by Hirst's watch. The Mur de la Côte was still before us, and on this the guide-books informed us two or three hundred steps were sometimes found necessary. If sixty steps cost an hour, what would be the cost of two hundred? The question was disheartening in the extreme, for the time at which we had calculated on reaching the summit was already passed, while the chief difficulties remained unconquered. Having hewn our way along the harder ice we reached snow. I again resorted to stamping to secure a footing, and while thus engaged became, for the first time, aware of the drain of force to which I was subjecting myself. The thought of being absolutely exhausted had never occurred to me, and from first to last I had taken no care to husband my strength. I always calculated that the *will* would serve me even should the muscles fail, but I now found that mechanical laws rule man in the long-run; that no effort of will, no power of spirit, can draw beyond a certain limit upon muscular force. The soul, it is true, can stir the body to action, but its function is to excite and apply force, and not to create it.

While stamping forward through the frozen crust I was compelled to pause at short intervals; then would set out again apparently fresh, to find, however, in a few minutes that my strength was gone, and that I required to rest once more. In this way I gained the summit of the Corridor, when Hirst came to the front, and I felt some relief in stepping slowly after him, making use of the holes into which his feet had sunk. He thus led the way to the base of the Mur de la Côte, the thought of which had so long cast a gloom upon us; here we left our rope behind us, and while pausing I asked Simond whether he did not feel a desire to go to the summit—

"*Bien sur*," was his reply, "*mais!*" Our guide's mind was so constituted that the "*mais*" seemed essential to its peace. I stretched my hand towards him, and said, "Simond, we must do it." One thing alone I felt could defeat us: the usual time of the ascent had been more than doubled, the day was already far spent, and if the ascent would throw our subsequent descent into night it could not be contemplated.

We now faced the Mur, which was by no means so bad as we had expected. Driving the iron claws of our boots into the scars made by the axe, and the spikes of our batons into the slope above our feet, we ascended steadily until the summit was attained, and the top of the mountain rose clearly above us. We congratulated ourselves upon this; but Simond, probably fearing that our joy might become too full, remarked, "*Mais le sommet est encore bien loin!*" It was, alas! too true. The snow became soft again, and our weary limbs sank in it as before. Our guide went on in front, audibly muttering his doubts as to our ability to reach the top, and at length he threw himself upon the snow, and exclaimed, "*Il faut le renoncer!*" Hirst now undertook the task of rekindling the guide's enthusiasm, after which Simond rose, exclaiming, "*Ah! comme ça me fait mal aux genoux*," and went forward. Two rocks break through the snow between the summit of the Mur and the top of the mountain; the first is called the Petits Mulets, and the highest the Derniers Rochers. At the former of these we paused to rest, and finished our scanty store of wine and provisions. We had not a bit of bread nor a drop of wine left; our brandy flasks were also nearly exhausted, and thus we had to contemplate the journey to the summit, and the subsequent descent to the Grands Mulets, without the slightest prospect of physical refreshment. The almost total loss of two nights' sleep, with two days' toil superadded, made me long for a few minutes' doze, so I stretched myself upon a composite couch of snow and granite, and immediately fell asleep. My friend, however, soon aroused me. "You quite frighten me," he said; "I have listened for some minutes, and have not heard you breathe once." I had, in reality, been taking deep draughts of the mountain air, but so silently as not to be heard.

I now filled our empty wine-bottle with snow and placed it in the sunshine, that we might have a little water on our return. We then rose; it was half-past two o'clock; we had been upwards of twelve hours climbing, and I calculated that, whether we reached the summit or not, we could at all events work *towards* it for another hour. To the sense of fatigue previously experienced, a new phenomenon was now added—the beating of the heart. We were incessantly pulled up by this, which sometimes became so intense as to suggest danger. I counted the number of paces which we were able to accomplish without resting, and found that at the end of every twenty, sometimes at the end of fifteen, we were compelled to pause. At each pause my heart throbbed audibly, as I leaned upon my staff, and the subsidence of this action was always the signal for further advance. My breathing was quick, but light and unimpeded. I endeavoured to ascertain whether the hip-joint, on account of the diminished atmospheric pressure, became loosened, so as to throw the weight of the leg upon the surrounding ligaments, but could not be certain about it. I also sought a little aid and encouragement from philosophy, endeavouring to remember what great things had been done by the accumulation of small quantities, and I urged upon myself that the present was a case in point, and that the summation of distances twenty paces each must finally place us at the top. Still the question of time left the matter long in doubt, and until we had passed the Derniers Rochers we worked on with the stern indifference of men who were doing their duty, and did not look to consequences. Here, however, a gleam of hope began to brighten our souls: the summit became visibly nearer, Simond showed more alacrity; at length success became certain, and at half-past three P.M. my friend and I clasped hands upon the top.

The summit of the mountain is an elongated ridge, which has been compared to the back of an ass. It was perfectly manifest that we were dominant over all other

First Ascent of Mont Blanc, 1857

mountains; as far as the eye could range Mont Blanc had no competitor. The summits which had looked down upon us in the morning were now far beneath us. The Dôme du Gouté, which had held its threatening *séracs* above us so long, was now at our feet. The Aiguille du Midi, Mont Blanc du Tacul, and the Monts Maudits, the Talèfre with its surrounding peaks, the Grand Jorasse, Mont Mallet, and the Aiguille du Géant, with our own familiar glaciers, were all below us. And as our eye ranged over the broad shoulders of the mountain, over ice hills and valleys, plateaux and far-stretching slopes of snow, the conception of its magnitude grew upon us, and impressed us more and more.

The clouds were very grand—grander indeed than anything I had ever before seen. Some of them seemed to hold thunder in their breasts, they were so dense and dark; others, with their faces turned sunward, shone with the dazzling whiteness of the mountain snow; while others again built themselves into forms resembling gigantic elm trees, loaded with foliage. Towards the horizon the luxury of colour added itself to the magnificent alternation of light and shade. Clear spaces of amber and ethereal green embraced the red and purple cumuli, and seemed to form the cradle in which they swung. Closer at hand squally mists, suddenly engendered, were driven hither and thither by local winds; while the clouds at a distance lay "like angels sleeping on the wing," with scarcely visible motion. Mingling with the clouds, and sometimes rising above them, were the highest mountain heads, and as our eyes wandered from peak to peak, onwards to the remote horizon, space itself seemed more vast from the manner in which the objects which it held were distributed.

I wished to repeat the remarkable experiment of De Saussure upon sound, and for this purpose had requested Simond to bring a pistol from Chamouni; but in the multitude of his cares he forgot it, and in lieu of it my host at the Montanvert had placed in two tin tubes, of the same size and shape, the same amount of gunpowder, securely closing the tubes afterwards, and furnishing each of them with a small lateral aperture. We now planted one of them upon the snow, and bringing a strip of amadou into communication with the touch-hole, ignited its most distant end: it failed; we tried again, and were successful, the explosion tearing asunder the little case which contained the powder. The sound was certainly not so great as I should have expected from an equal quantity of powder at the sea level.[1]

The snow upon the summit was indurated, but of an exceedingly fine grain, and the beautiful effect already referred to as noticed upon the Stelvio was strikingly manifest. The hole made by driving the baton into the snow was filled with a delicate blue light; and, by management, its complementary pinky yellow could also be produced. Even the iron spike at the end of the baton made a hole sufficiently deep to exhibit the blue colour, which certainly depends on the size and arrangement of the snow crystals. The firmament above us was without a cloud, and of a darkness almost equal to that which surrounded the moon at 2 A.M. Still, though the sun was shining, a breeze, whose tooth had been sharpened by its passage over the snow-fields, searched us through and through. The day was also waning, and, urged by the warnings of our ever prudent guide, we at length began the descent.

Gravity was now in our favour, but gravity could not entirely spare our wearied limbs, and where we sank in the snow we found our downward progress very trying. I suffered from thirst, but after we had divided the liquefied snow at the Petits Mulets amongst us we had nothing to drink. I crammed the clean snow into my mouth, but the process of melting was slow and tantalising to a parched throat, while the chill was painful to the teeth.

---

[1] I fired the second case in a field in Hampshire, and, as far as my memory enabled me to make the comparison, found its sound considerably *denser*, if I may use the expression. In 1859 I had a pistol fired at the summit of Mont Blanc: its sound was sensibly feebler and *shorter* than in the valley; it resembled somewhat the discharge of a cork from a champagne bottle, though much louder, but it could not be at all compared to the sound of a common cracker.

We marched along the Corridor, and crossed cautiously the perilous slope on which we had cut steps in the morning, breathing more freely after we had cleared the ice-precipice before described. Along the base of this precipice we now wound, diverging from our morning's track, in order to get surer footing in the snow; it was like flour, and while descending to the Grand Plateau we sometimes sank in it nearly to the waist. When I endeavoured to squeeze it, so as to fill my flask, it at first refused to cling together, behaving like so much salt; the heat of the hand, however, soon rendered it a little moist, and capable of being pressed into compact masses. The sun met us here with extraordinary power; the heat relaxed my muscles, but when fairly immersed in the shadow of the Dôme du Gouté, the coolness restored my strength, which augmented as the evening advanced. Simond insisted on the necessity of haste, to save us from the perils of darkness. "*On peut perir*" was his repeated admonition, and he was quite right. We reached the region of *ponts*, more weary, but, in compensation, more callous, than we had been in the morning, and moved over the soft snow of the bridges as if we had been walking upon eggs. The valley of Chamouni was filled with brown-red clouds, which crept towards us up the mountain; the air around and above us was, however, clear, and the chastened light told us that day was departing. Once as we hung upon a steep slope, where the snow was exceedingly soft, Hirst omitted to make his footing sure; the soft mass gave way, and he fell, uttering a startled shout as he went down the declivity. I was attached to him, and, fixing my feet suddenly in the snow, endeavoured to check his fall, but I seemed a mere feather in opposition to the force with which he descended.[1] I fell, and went down after him; and we carried quite an avalanche of snow along with us, in which we were almost completely hidden at the bottom of the slope. All further dangers, however, were soon past, and we went at a headlong speed to the base of the

---

[1] I believe that I could stop him now (1860).

Grands Mulets; the sound of our batons against the rocks calling Huxley forth. A position more desolate than his had been can hardly be imagined. For seventeen hours he had been there. He had expected us at two o'clock in the afternoon; the hours came and passed, and till seven in the evening he had looked for us. "To the end of my life," he said, "I shall never forget the sound of those batons." It was his turn now to nurse me, which he did, repaying my previous care of him with high interest. We were all soon stretched, and, in spite of cold and hard boards, I slept at intervals; but the night, on the whole, was a weary one, and we rose next morning with muscles more tired than when we lay down.

Friday, 14th August.—Hirst was almost blind this morning; and our guide's eyes were also greatly inflamed. We gathered our things together, and bade the Grands Mulets farewell. It had frozen hard during the night, and this, on the steeper slopes, rendered the footing very insecure. Simond, moreover, appeared to be a little bewildered, and I sometimes preceded him in cutting the steps, while Hirst moved among the crevasses like a blind man; one of us keeping near him, so that he might feel for the actual places where our feet had rested, and place his own in the same position. It cost us three hours to cross from the Grands Mulets to the Pierre l'Echelle, where we discarded our leggings, had a mouthful of food, and a brief rest. Once upon the safe earth Simond's powers seemed to be restored, and he led us swiftly downwards to the little auberge beside the Cascade du Tard, where we had some excellent lemonade, equally choice cognac, fresh strawberries and cream. How sweet they were, and how beautiful we thought the peasant girl who served them! Our guide kept a little hotel, at which we halted, and found it clean and comfortable. We were, in fact, totally unfit to go elsewhere. My coat was torn, holes were kicked through my boots, and I was altogether ragged and shabby. A warm bath before dinner refreshed all mightily. Dense clouds now lowered upon Mont Blanc, and we had not been an hour at

Chamouni when the breaking up of the weather was announced by a thunder-peal. We had accomplished our journey just in time.

XII

After our return we spent every available hour upon the ice, working at questions which shall be treated under their proper heads, each day's work being wound up by an evening of perfect enjoyment. Roast mutton and fried potatoes were our incessant fare, for which, after a little longing for a change at first, we contracted a final and permament love. As the year advanced, moreover, and the grass sprouted with augmented vigour on the slopes of the Montanvert, the mutton, as predicted by our host, became more tender and juicy. We had also some capital Sallenches beer, cold as the glacier water, but effervescent as champagne. Such were our food and drink. After dinner we gathered round the pine-fire, and I can hardly think it possible for three men to be more happy than we then were. It was not the goodness of the conversation, nor any high intellectual element, which gave the charm to our gatherings; the gladness grew naturally out of our own perfect health, and out of the circumstances of our position. Every fibre seemed a repository of latent joy, which the slightest stimulus sufficed to bring into conscious action.

On the 17th I penetrated with Simond through thick gloom to the Tacul; on the 18th we set stakes at the same place: on the same day, while crossing the medial moraine of the Talèfre, a little below the cascade, a singular noise attracted my attention; it seemed at first as if a snake were hissing about my feet. On changing my position the sound suddenly ceased, but it soon recommenced. There was some snow upon the glacier, which I removed, and placed my ear close to the ice, but it was difficult to fix on the precise spot from which the sound issued. I cut away the disintegrated portion of the surface, and at length discovered a minute crack, from which a stream of air issued, which I could feel as a cold blast against my hand. While cutting away the surface further, I stopped the little "blower." A marmot screamed near me, and while I paused to look at the creature scampering up the crags, the sound commenced again, changing its note variously—hissing like a snake, singing like a kettle, and sometimes chirruping intermittently like a bird. On passing my fingers to and fro over the crack, I obtained a succession of audible puffs; the current was sufficiently strong to blow away the corner of my gauze veil when held over the fissure. Still the crack was not wide enough to permit of the entrance of my finger-nail; and to issue with such force from so minute a rent the air must have been under considerable pressure. The origin of the blower was in all probability the following:—When the ice is recompacted after having descended a cascade, it is next to certain that chambers of air will be here and there enclosed, which, being powerfully squeezed afterwards, will issue in the manner described whenever a crack in the ice furnishes it with a means of escape. In my experiments on flowing mud, for example, the air entrapped in the mass while descending from the sluice into the trough, bursts in bubbles from the surface at a short distance downwards.

I afterwards examined the Talèfre cascade from summit to base, with reference to the structure, until at the close of the day thickening clouds warned me off. I went down the glacier at a trot, guided by the boulders capped with little cairns which marked the route. The track which I had pursued for the last five weeks amid the crevasses near l'Angle was this day barely passable. The glacier had changed, my work was drawing to a close, and, as I looked at the objects which had now become so familiar to me, I felt that, though not viscous, the ice did not lack the quality of "adhesiveness," and I felt a little sad at the thought of bidding it so soon farewell.

At some distance below the Montanvert the Mer de Glace is riven from side to side by transverse crevasses: these fissures indicate that the glacier where they occur is in a state of longitudinal strain which produces transverse fracture. I wished to ascertain the amount of stretching which the glacier here demanded, and which the ice was not able to give; and for this purpose desired to compare the velocity of a line set out across the fissured portion with that of a second line staked out across the ice before it had become thus fissured. A previous inspection of the glacier through the telescope of our theodolite induced us to fix on a place which, though much riven, still did not exclude the hope of our being able to reach the other side. Each of us was, as usual, armed with his own axe; and carrying with us suitable stakes, my guide and myself entered upon this portion of the glacier on the morning of the 19th of August.

I was surprised on entering to find some veins of white ice, which from their position and aspect appeared to be derived from the Glacier du Géant; but to these I shall subsequently refer. Our work was extremely difficult; we penetrated to some distance along one line, but were finally forced back, and compelled to try another. Right and left of us were profound fissures, and once a cone of ice forty feet high leaned quite over our track. In front of us was a second leaning mass borne by a mere stalk, and so topheavy that one wondered why the slight pedestal on which it rested did not suddenly crack across. We worked slowly forwards, and soon found ourselves in the shadow of the topheavy mass above referred to; and from which I escaped with a wounded hand, caused by over-haste. Simond surmounted the next ridge and exclaimed, "*Nous nous trouverons perdus!*" I reached his side, and on looking round the place saw that there was no footing for man. The glacier here was cut up into thin wedges, separated from each other by profound chasms, and the wedges were so broken across as to render creeping along their edges quite impossible. Thus brought to a stand, I fixed a stake at the point where we were forced to halt, and retreated along edges of detestable granular ice, which fell in showers into the crevasses when struck by the axe. At one place an exceedingly deep fissure was at our left, which was joined, at a sharp angle, by another at our right, and we were compelled to cross at the place of intersection: to do this we had to trust ourselves to a projecting knob of that vile rotten ice which I had learned to fear since my experience of it on the Col du Géant. We finally escaped, and set out our line at another place, where the glacier, though badly cut, was not impassable.

On the 20th we made a series of final measurements at the Tacul, and determined the motion of two lines which we had set out upon the previous day. On the 21st we quitted the Montanvert; I had been there from the 15th of July, and the longer I remained the better I liked the establishment and the people connected with it. It was then managed by Joseph Tairraz and Jules Charlet, both of whom showed us every attention. In 1858 and 1859 I had occasion to revisit the establishment, which was then managed by Jules and his brother, and found in it the same good qualities. During my winter expedition of 1859 I also found the same readiness to assist me in every possible way; honest Jules expressing his willingness to ascend through the snow to the auberge if I thought his presence would in any degree contribute to my comfort.

We crossed the glacier, and descended by the Chapeau to the Cascade des Bois, the inclination of which and of the lower portion of the glacier we then determined. The day was magnificent. Looking upwards, the Aiguilles de Charmoz and du Dru rose right and left like sentinels of the valley, while in front of us the ice descended the steep, a bewildering mass of crags and chasms. At the other side was the pine-clad slope of the Montanvert. Further on the Aiguille du Midi threw its granite pyramid between us and Mont Blanc; on the Dôme du Gouté the *séracs* of the mountain were to be seen, while issuing as if from a cleft in the mountain side the Glacier des Bossons thrust through the black pines its snowy tongue. Below us was the beautiful valley of Chamouni itself, through which the Arve and Arveiron rushed like enlivening spirits. We finally examined a grand old moraine produced by a Mer de Glace of other ages, when the ice quite crossed the valley of Chamouni and abutted against the opposite mountain-wall.

Simond had proved himself a very valuable assistant: he was intelligent and perfectly trustworthy; and though the peculiar nature of my work sometimes caused me to attempt things against which his prudence protested, he lacked neither strength nor courage. On reaching Chamouni and adding up our accounts, I found that I had not sufficient cash to pay him; money was waiting for me at the post-office in Geneva, and thither it was arranged that my friend Hirst should proceed next morning, while I was to await the arrival of the money at Chamouni. My guide heard of this arrangement, and divined its cause: he came to me, and in the most affectionate manner begged of me to accept from him the loan of 500 francs. Though I did not need the loan, the mode in which it was offered to me augmented the kindly feelings which I had long entertained towards Simond, and I may add that my intercourse with him since has served merely to confirm my first estimate of his worthiness.

EXPEDITION OF 1858

XIII

I HAD confined myself during the summer of 1857 to the Mer de Glace and its tributaries, desirous to make my knowledge accurate rather than extensive. I had made the acquaintance of all accessible parts of the glacier, and spared no pains to master both the details and the meaning of the laminated structure of the ice, but I found no fact upon which I could take my stand and say to an advocate of an opposing theory, "This is unassailable." In experimental science we have usually the power of changing the conditions at pleasure; if Nature does not reply to a question we throw it into another form; a combining of conditions is, in fact, the essence of experiment. To meet the requirements of the present question, I could not twist the same glacier into various shapes, and throw it into different states of strain and pressure; but I might, by visiting many glaciers, find all needful conditions fulfilled in detail, and by observing these I hoped to confer upon the subject the character and precision of a true experimental inquiry.

The summer of 1858 was accordingly devoted to this purpose, when I had the good fortune to be accompanied by Professor Ramsay, the author of some extremely interesting papers upon ancient glaciers. Taking Zürich, Schaffhausen, and Lucerne in our way, we crossed the Brünig on the 22nd of July, and met my guide, Christian Lauener, at Meyringen. On the 23rd we visited the glacier of Rosenlaui, and the glacier of the Schwartzwald, and reached Grindelwald in the evening of the same day. My expedition with Mr. Huxley had taught me that the Lower Grindelwald Glacier was extremely instructive, and I was anxious to see many parts of it once more; this I did, in company with Ramsay, and we also spent a day upon the upper glacier, after which our path lay over the Strahleck to the glaciers of the Aar and of the Rhone.

PASSAGE OF THE STRAHLECK

XIV

On Monday, the 26th of July, we were called at 4 A.M., and found the weather very unpromising, but the two mornings which preceded it had also been threatening without any evil result. There was, it is true, something more than usually hostile in the aspect of the clouds which sailed sullenly from the west, and smeared the air and mountains as if with the dirty smoke of a manufacturing town. We despatched our coffee, went down to the bottom of the Grindelwald valley, up the opposite slope, and were soon amid the gloom of the pines which partially cover it. On emerging from these, a watery gleam on the mottled head of the Eiger was the only evidence of direct sunlight in that direction. To our left was the Wetterhorn surrounded by wild and disorderly clouds, through the fissures of which the morning light glared strangely. For a time the Heisse Platte was seen, a dark brown patch amid the ghastly blue which overspread the surrounding slopes of snow. The clouds once rolled up, and revealed for a moment the summits of the Viescherhörner; but they immediately settled down again, and hid the mountains from top to base. Soon afterwards they drew themselves partially aside, and a patch of blue over the Strahleck gave us hope and pleasure. As we ascended, the prospect in front of us grew better, but that behind us—and the wind came from behind—grew worse. Slowly and stealthily the dense neutral-tint masses crept along the sides of the mountains, and seemed to dog us like spies; while over the glacier hung a thin veil of fog, through which gleamed the white minarets of the ice.

When we first spoke of crossing the Strahleck, Lauener

said it would be necessary to take two guides at least; but after a day's performance on the ice he thought we might manage very well by taking, in addition to himself, the herd of the alp, over the more difficult part of the pass. He had further experience of us on the second day, and now, as we approached the herd's hut, I was amused to hear him say that he thought any assistance beside his own unnecessary. Relying upon ourselves, therefore, we continued our route, and were soon upon the glacier, which had been rendered smooth and slippery through the removal of its disintegrated surface by the warm air. Crossing the Strahleck branch of the glacier to its left side, we climbed the rocks to the grass and flowers which clothe the slopes above them. Our way sometimes lay over these, sometimes along the beds of streams, across turbulent brooks, and once around the face of a cliff, which afforded us about an inch of ledge to stand upon, and some protruding splinters to lay hold of by the hands. Having reached a promontory which commanded a fine view of the glacier, and of the ice cascade by which it was fed, I halted, to check the observations already made from the side of the opposite mountain. Here, as there, cliffy ridges were seen crossing the cascade of the glacier, with interposed spaces of dirt and débris—the former being toned down, and the latter squeezed towards the base of the fall, until finally the ridges swept across the glacier, in gentle swellings, from side to side; while the valleys between them, holding the principal share of the superficial impurity, formed the cradles of the so-called Dirt-Bands. These swept concentric with the protuberances across the glacier, and remained upon its surface even after the swellings had disappeared. The swifter flow of the centre of the glacier tends of course incessantly to lengthen the loops of the bands, and to thrust the summits of the curves which they form more and more in advance of their lateral portions. The depressions between the protuberances appeared to be furrowed by minor wrinkles, as if the ice of the depressions had yielded more than that of the protuberances. This, I think, is extremely probable,

Passage of the Strahleck

though it has never yet been proved. Three stakes, placed, one on the summit, another on the frontal slope, and another at the base of a protuberance, would, I think, move with unequal velocities. They would, I think, show that, upon the large and general motion of the glacier, smaller motions are superposed, as minor oscillations are known to cover parasitically the large ones of a vibrating string. Possibly, also, the dirt-bands may owe something to the squeezing of impurities out of the glacier to its surface in the intervals between the swellings. From our present position we could also see the swellings on the Viescherhörner branch of the glacier, in the valleys between which coarse shingle and débris were collected, which would form dirt-bands if they could. On neither branch, however, do the bands attain the definition and beauty which they possess upon the Mer de Glace.

After an instructive lesson we faced our task once more, passing amid crags and boulders, and over steep moraines, from which the stones rolled down upon the slightest disturbance. While crossing a slope of snow with an inclination of 45°, my footing gave way, I fell, but turned promptly on my face, dug my staff deeply into the snow, and arrested the motion before I had slid a dozen yards. Ramsay was behind me, speculating whether he should be able to pass the same point without slipping; before he reached it, however, the snow yielded, he fell, and slid swiftly downwards. Lauener, whose attention had been aroused by my fall, chanced to be looking round when Ramsay's footing yielded. With the velocity of a projectile he threw himself upon my companion, seized him, and brought him to rest before he had reached the bottom of the slope. The act made a very favourable impression upon me, it was so prompt and instinctive. An eagle could not swoop upon its prey with more directness of aim and swiftness of execution.

While this went on the clouds were playing hide and seek with the mountains. The ice-crags and pinnacles to our left, looming through the haze, seemed of gigantic proportions, reminding one of the Hades of Byron's 'Cain.'

"How sunless and how vast are these dim realms!"

We climbed for some time along the moraine which flanks the cascade, and on reaching the level of the brow Lauener paused, cast off his knapsack, and declared for breakfast. While engaged with it the dense clouds which had crammed the gorge and obscured the mountains, all melted away, and a scene of indescribable magnificence was revealed. Overhead the sky suddenly deepened to dark blue, and against it the Finsteraarhorn projected his dark and mighty mass. Brown spurs jutted from the mountain, and between them were precipitous snow-slopes, fluted by the descent of rocks and avalanches, and broken into ice-precipices lower down. Right in front of us, and from its proximity more gigantic to the eye, was the Shreckhorn, while from couloirs and mountain-slopes the matter of glaciers yet to be was poured into the vast basin on the rim of which we now stood.

This it was next our object to cross; our way lying in part through deep snow-slush, the scene changing perpetually from blue heaven to grey haze which massed itself at intervals in dense clouds about the mountains. After crossing the basin our way lay partly over slopes of snow, partly over loose shingle, and at one place along the edge of a formidable precipice of rock. We sat down sometimes to rest, and during these pauses, though they were very brief, the scene had time to go through several of its Protean mutations. At one moment all would be perfectly serene, no cloud in the transparent air to tell us that any portion of it was in motion, while the blue heaven threw its flattened arch over the magnificent amphitheatre. Then in an instant, from some local cauldron, the vapour would boil up suddenly, eddying wildly in the air, which a moment before seemed so still, and enveloping the entire scene. Thus the space enclosed by the Finsteraarhorn, the Viescherhörner, and the Shreckhorn, would at one moment be filled with fog to the mountain heads, every trace of which a few minutes

Passage of the Strahleck

sufficed to sweep away, leaving the unstained blue of heaven behind it, and the mountains showing sharp and jagged outlines in the glassy air. One might be almost led to imagine that the vapour molecules endured a strain similar to that of water cooled below its freezing point, or heated beyond its boiling point; and that, on the strain being relieved by the sudden yielding of the opposing force, the particles rushed together, and thus filled in an instant the clear atmosphere with aqueous precipitation.

I had no idea that the Strahleck was so fine a pass. Whether it is the quality of my mind to take in the glory of the present so intensely as to make me forgetful of the glory of the past, I know not, but it appeared to me that I had never seen anything finer than the scene from the summit. The amphitheatre formed by the mountains seemed to me of exceeding magnificence; nor do I think that my feeling was subjective merely; for the simple magnitude of the masses which built up the spectacle would be sufficient to declare its grandeur. Looking down towards the glacier of the Aar, a scene of wild beauty and desolation presented itself. Not a trace of vegetation could be seen along the whole range of the bounding mountains; glaciers streamed from their shoulders into the valley beneath, where they welded themselves to form the Finsteraar affluent of the Unteraar Glacier.

After a brief pause, Lauener again strapped on his knapsack, and tempered both will and muscles by the remark that our worst piece of work was now before us. From the place where we sat, the mountain fell precipitously for several hundred feet; and down the weathered crags, and over the loose shingle which encumbered their ledges, our route now lay. Lauener was in front, cool and collected, lending at times a hand to Ramsay, and a word of encouragement to both of us, while I brought up the rear. I found my full haversack so inconvenient that I once or twice thought of sending it down the crags in advance of me, but Lauener assured me that it would be utterly destroyed before reaching the bottom. My complaint against it was, that at critical places it sometimes came between me and the face of the cliff, pushing me away from the latter so as to throw my centre of gravity almost beyond the base intended to support it. We came at length upon a snow-slope, which had for a time an inclination of 50°; then once more to the rocks; again to the snow, which was both steep and deep. Our batons were at least six feet long: we drove them into the snow to secure an anchorage, but they sank to their very ends, and we merely retained a length of them sufficient for a grasp. This slope was intersected by a so-called Bergschrund, the lower portion of the slope being torn away from its upper portion so as to form a crevasse that extended quite round the head of the valley. We reached its upper edge; the chasm was partially filled with snow, which brought its edges so near that we cleared it by a jump. The rest of the slope was descended by a *glissade*. Each sat down upon the snow, and the motion, once commenced, swiftly augmented to the rate of an avalanche, and brought us pleasantly to the bottom.

As we looked from the heights, we could see that the valley through which our route lay was filled with grey fog: into this we soon plunged, and through it we made our way towards the Abschwung. The inclination of the glacier was our only guide, for we could see nothing. Reaching the confluence of the Finsteraar and Lauteraar branches, we went downwards with long swinging strides, close alongside the medial moraine of the trunk glacier. The glory of the morning had its check in the dull gloom of the evening. Across streams, amid dirt-cones and glacier-tables, and over the long reach of shingle which covers the end of the glacier, we plodded doggedly, and reached the Grimsel at 7 P.M., the journey having cost a little more than fourteen hours.

XV

We made the Grimsel our station for a day, which was spent in examining the evidences of ancient glacier action in the valley of Hasli. Near the Hospice, but at the

Passage of the Strahleck 89

opposite side of the Aar, rises a mountain-wall of hard granite, on which the flutings and groovings are magnificently preserved. After a little practice the eye can trace with the utmost precision the line which marks the level of the ancient ice: above this the crags are sharp and rugged; while below it the mighty grinder has rubbed off the pinnacles of the rocks and worn their edges away. The height to which this action extends must be nearly two thousand feet above the bed of the present valley. It is also easy to see the depth to which the river has worked its channel into the ancient rocks. In some cases the road from Guttanen to the Grimsel lay right over the polished rocks, asperities being supplied by the chisel of man in order to prevent travellers from slipping on their slopes. Here and there also huge protuberant crags were rounded into domes almost as perfect as if chiselled by art. To both my companion and myself this walk was full of instruction and delight.

On the 28th of July we crossed the Grimsel Pass, and traced the scratchings to the very top of it. Ramsay remarked that their direction changed high up the pass, as if a tributary from the summit had produced them, while lower down they merged into the general direction of the glacier which had filled the principal valley. From the summit of the Mayenwand we had a clear view of the glacier of the Rhone; and to see the lower portion of this glacier to advantage no better position can be chosen. The dislocation of its cascade, the spreading out of the ice below, its system of radial crevasses, and the transverse sweep of its structural groovings, may all be seen. A few hours afterwards we were amid the wild chasms at the brow of the ice-fall, where we worked our way to the centre of the ice, but were unable to attain the opposite side.

Having examined the glacier both above and below the cascade, we went down the valley to Viesch, and ascended thence, on the 30th of July, to the Hôtel Jungfrau on the slopes of the Æggischhorn. On the following day we climbed to the summit of the mountain, and from a sheltered nook enjoyed the glorious prospect which it commands. The wind was strong, and fleecy clouds flew over the heavens; some of which, as they formed and dispersed themselves about the flanks of the Aletschhorn, showed extraordinary iridescences.

The sunbeams called us early on the morning of the 1st of August. No cloud rested on the opposite range of the Valais mountains, but on looking towards the Æggischhorn we found a cap upon its crest; we looked again—the cap had disappeared and a serene heaven stretched overhead. As we breasted the alp the moon was still in the sky, paling more and more before the advancing day; a single hawk swung in the atmosphere above us; clear streams babbled from the hills, the louder sounds reposing on a base of music; while groups of cows with tinkling bells browsed upon the green alp. Here and there the grass was dispossessed, and the flanks of the mountain were covered by the blocks which had been cast down from the summit. On reaching the plateau at the base of the final pyramid, we rounded the mountain to the right and came over the lonely and beautiful Märjelen See. No doubt the hollow which this lake fills had been scooped out in former ages by a branch of the Aletsch Glacier; but long ago the blue ice gave place to blue water. The glacier bounds it at one side by a vertical wall of ice sixty feet in height: this is incessantly undermined, a roof of crystal being formed over the water, till at length the projecting mass, becoming too heavy for its own rigidity, breaks and tumbles into the lake. Here, attacked by sun and air, its blue surface is rendered dazzlingly white, and several icebergs of this kind now floated in the sunlight; the water was of a glassy smoothness, and in its blue depths each ice mass doubled itself by reflection.[1]

The Aletsch is the grandest glacier in the Alps: over it we now stood, while the bounding mountains poured vast feeders into the noble stream. The Jungfrau was in front of us without a cloud, and apparently so near that I

---

[1] A painting of this exquisite lake has been recently executed by Mr. George Barnard.

Passage of the Strahleck

proposed to my guide to try it without further preparation. He was enthusiastic at first, but caution afterwards got the better of his courage. At some distance up the glacier the snow-line was distinctly drawn, and from its edge upwards the mighty shoulders of the hills were heavy laden with the still powdery material of the glacier.

Amid blocks and débris we descended to the ice: the portion of it which bounded the lake had been sapped, and a space of a foot existed between ice and water: numerous chasms were formed here, the mass being thus broken, preparatory to being sent adrift upon the lake. We crossed the glacier to its centre, and looking down it the grand peaks of the Mischabel, the noble cone of the Weisshorn, and the dark and stern obelisk of the Matterhorn, formed a splendid picture. Looking upwards, a series of most singularly contorted dirt-bands revealed themselves upon the surface of the ice. I sought to trace them to their origin, but was frustrated by the snow which overspread the upper portion of the glacier. Along this we marched for three hours, and came at length to the junction of the four tributary valleys which pour their frozen streams into the great trunk valley. The glory of the day, and that joy of heart which perfect health confers, may have contributed to produce the impression, but I thought I had never seen anything to rival in magnificence the region in the heart of which we now found ourselves. We climbed the mountain on the right-hand side of the glacier, where, seated amid the riven and weather-worn crags, we fed our souls for hours on the transcendent beauty of the scene.

We afterwards redescended to the glacier, which at this place was intersected by large transverse crevasses, many of which were apparently filled with snow, while over others a thin and treacherous roof was thrown. In some cases the roof had broken away, and revealed rows of icicles of great length and transparency pendent from the edges. We at length turned our faces homewards, and looking down the glacier I saw at a great distance something moving on the ice. I first thought it was a man, though it seemed strange that a man should be there alone. On drawing my guide's attention to it he at once pronounced it to be a chamois, and I with my telescope immediately verified his statement. The creature bounded up the glacier at intervals, and sometimes the vigour of its spring showed that it had projected itself over a crevasse. It approached us sometimes at full gallop: then would stop, look toward us, pipe loudly, and commence its race once more. It evidently made the reciprocal mistake to my own, imagining us to be of its own kith and kin. We sat down upon the ice the better to conceal our forms, and to its whistle our guide whistled in reply. A joyous rush was the creature's first response to the signal; but it afterwards began to doubt, and its pauses became more frequent. Its form at times was extremely graceful, the head erect in the air, its apparent uprightness being augmented by the curvature which threw its horns back. I watched the animal through my glass until I could see the glistening of its eyes; but soon afterwards it made a final pause, assured itself of its error, and flew with the speed of the wind to its refuge in the mountains.

ASCENT OF THE FINSTERAARHORN, 1858

XVI

Since my arrival at the hotel on the 30th of July I had once or twice spoken about ascending the Finsteraarhorn, and on the 2nd of August my host advised me to avail myself of the promising weather. A guide, named Bennen, was attached to the hotel, a remarkable-looking man, between thirty and forty years old, of middle stature, but very strongly built. His countenance was frank and firm, while a light of good-nature at times twinkled in his eye. Altogether the man gave me the impression of physical strength, combined with decision of character. The proprietor had spoken to me many times of the strength and courage of this man, winding up

Ascent of the Finsteraarhorn, 1858

his praises of him by the assurance that if I were killed in Bennen's company there would be two lives lost, for that the guide would assuredly sacrifice himself in the effort to save his *Herr*.

He was called, and I asked him whether he would accompany me alone to the top of the Finsteraarhorn. To this he at first objected, urging the possibility of his having to render me assistance, and the great amount of labour which this might entail upon him; but this was overruled by my engaging to follow where he led, without asking him to render me any help whatever. He then agreed to make the trial, stipulating, however, that he should not have much to carry to the cave of the Faulberg, where we were to spend the night. To this I cordially agreed, and sent on blankets, provisions, wood, and hay, by two porters.

My desire, in part, was to make a series of observations at the summit of the mountain, while a similar series was made by Professor Ramsay in the valley of the Rhone, near Viesch, with a view to ascertaining the permeability of the lower strata of the atmosphere to the radiant heat of the sun. During the forenoon of the 2nd I occupied myself with my instruments, and made the proper arrangements with Ramsay. I tested a mountain-thermometer which Mr. Casella had kindly lent me, and found the boiling-point of water on the dining-room table of the hotel to be 199.29° Fahrenheit. At about three o'clock in the afternoon we quitted the hotel, and proceeded leisurely with our two guides up the slope of the Æggischhorn. We once caught a sight of the topmost pinnacle of the Finsteraarhorn; beside it was the Rothhorn, and near this again the Oberaarhorn, with the Viescher Glacier streaming from its shoulders. On the opposite side we could see, over an oblique buttress of the mountain on which we stood, the snowy summit of the Weisshorn; to the left of this was the ever grim and lonely Matterhorn; and farther to the left, with its numerous snowcones, each with its attendant shadow, rose the mighty Mischabel. We descended, and crossed the stream which flows from the Märjelen See, into which a large mass of the glacier had recently fallen, and was now afloat as an iceberg. We passed along the margin of the lake, and at the junction of water and ice I bade Ramsay good-bye. At the commencement of our journey upon the ice, whenever we crossed a crevasse, I noticed Bennen watching me; his vigilance, however, soon diminished, whence I gathered that he finally concluded that I was able to take care of myself. Clouds hovered in the atmosphere throughout the whole time of our ascent; one smoky-looking mass marred the glory of the sunset, but at some distance was another which exhibited colours almost as rich and varied as those of the solar spectrum. I took the glorious banner thus unfurled as a sign of hope, to check the despondency which its gloomy neighbour was calculated to produce.

Two hours' walking brought us near our place of rest; the porters had already reached it, and were now returning. We deviated to the right, and, having crossed some ice-ravines, reached the lateral moraine of the glacier, and picked our way between it and the adjacent mountain-wall. We then reached a kind of amphitheatre, crossed it, and climbing the opposite slope, came to a triple grotto formed by clefts in the mountain. In one of these a pine-fire was soon blazing briskly, and casting its red light upon the surrounding objects, though but half dispelling the gloom from the deeper portions of the cell. I left the grotto, and climbed the rocks above it to look at the heavens. The sun had quitted our firmament, but still tinted the clouds with red and purple; while one peak of snow in particular glowed like fire, so vivid was its illumination. During our journey upwards the Jungfrau never once showed her head, but, as if in ill temper, had wrapped her vapoury veil around her. She now looked more good-humoured, but still she did not quite remove her hood; though all the other summits, without a trace of cloud to mask their beautiful forms, pointed heavenward. The calmness was perfect; no sound of living creature, no whisper of a breeze, no gurgle of water, no rustle of débris, to break the deep and solemn silence. Surely, if beauty be an object of worship, those glorious mountains, with rounded shoulders of the purest white—snow-crested and star-gemmed—were well calculated to excite sentiments of adoration.

I returned to the grotto, where supper was prepared and waiting for me. The boiling-point of water, at the level of the "kitchen" floor, I found to be 196° Fahr. Nothing could be more picturesque than the aspect of the cave before we went to rest. The fire was gleaming ruddily. I sat upon a stone bench beside it, while Bennen was in front with the red light glimmering fitfully over him. My boiling-water apparatus, which had just been used, was in the foreground; and telescopes, opera-glasses, haversacks, wine-keg, bottles, and mattocks, lay confusedly around. The heavens continued to grow clearer, the thin clouds, which had partially overspread the sky, melting gradually away. The grotto was comfortable; the hay sufficient materially to modify the hardness of the rock, and my position at least sheltered and warm. One possibility remained that might prevent me from sleeping—the snoring of my companion; he assured me, however, that he did not snore, and we lay down side by side. The good fellow took care that I should not be chilled; he gave me the best place, by far the best part of the clothes, and may have suffered himself in consequence; but, happily for him, he was soon oblivious of this. Physiologists, I believe, have discovered that it is chiefly during sleep that the muscles are repaired; and ere long the sound I dreaded announced to me at once the repair of Bennen's muscles and the doom of my own. The hollow cave resounded to the deep-drawn snore. I once or twice stirred the sleeper, breaking thereby the continuity of the phenomenon; but it instantly pieced itself together again, and went on as before. I had not the heart to wake him, for I knew that upon him would devolve the chief labour of the coming day. At half-past one he rose and prepared coffee, and at two o'clock I was engaged upon the beverage. We afterwards packed up our provisions and instruments. Bennen bore the former, I the latter, and at three o'clock we set out.

We first descended a steep slope to the glacier, along

which we walked for a time. A spur of the Faulberg jutted out between us and the ice-laden valley through which we must pass; this we crossed in order to shorten our way and to avoid crevasses. Loose shingle and boulders overlaid the mountain; and here and there walls of rock opposed our progress, and rendered the route far from agreeable. We then descended to the Grünhorn tributary, which joins the trunk glacier at nearly a right angle, being terminated by a saddle which stretches across from mountain to mountain, with a curvature as graceful and as perfect as if drawn by the instrument of a mathematician. The unclouded moon was shining, and the Jungfrau was before us so pure and beautiful, that the thought of visiting the "Maiden" without further preparation occurred to me. I turned to Bennen, and said, "Shall we try the Jungfrau?" I think he liked the idea well enough, though he cautiously avoided incurring any responsibility. "If you desire it, I am ready," was his reply. He had never made the ascent, and nobody knew anything of the state of the snow this year; but Lauener had examined it through a telescope on the previous day, and pronounced it dangerous. In every ascent of the mountain hitherto made, ladders had been found indispensable, but we had none. I questioned Bennen as to what he thought of the probabilities, and tried to extract some direct encouragement from him; but he said that the decision rested altogether with myself, and it was his business to endeavour to carry out that decision. "We will attempt it, then," I said, and for some time we actually walked towards the Jungfrau. A grey cloud drew itself across her summit, and clung there. I asked myself why I deviated from my original intention? The Finsteraarhorn was higher, and therefore better suited for the contemplated observations. I could in no wise justify the change, and finally expressed my scruples. A moment's further conversation caused us to "right about," and front the saddle of the Grünhorn.

The dawn advanced. The eastern sky became illuminated and warm, and high in the air across the ridge in front of us stretched a tongue of cloud like a red flame, and equally fervid in its hue. Looking across the trunk glacier, a valley which is terminated by the Lötsch saddle was seen in a straight line with our route, and I often turned to look along this magnificent corridor. The mightiest mountains in the Oberland form its sides; still, the impression which it makes is not that of vastness or sublimity, but of loveliness not to be described. The sun had not yet smitten the snows of the bounding mountains, but the saddle carved out a segment of the heavens which formed a background of unspeakable beauty. Over the rim of the saddle the sky was deep orange, passing upwards through amber, yellow, and vague ethereal green to the ordinary firmamental blue. Right above the snow-curve purple clouds hung perfectly motionless, giving depth to the spaces between them. There was something saintly in the scene. Anything more exquisite I had never beheld.

We marched upwards over the smooth crisp snow to the crest of the saddle, and here I turned to take a last look along that grand corridor, and at that wonderful "daffodil sky." The sun's rays had already smitten the snows of the Aletschhorn; the radiance seemed to infuse a principle of life and activity into the mountains and glaciers, but still that holy light shone forth, and those motionless clouds floated beyond, reminding one of that eastern religion whose essence is the repression of all action and the substitution for it of immortal calm. The Finsteraarhorn now fronted us; but clouds turbaned the head of the giant, and hid it from our view. The wind, however, being north, inspired us with a strong hope that they would melt as the day advanced. I have hardly seen a finer ice-field than that which now lay before us. Considering the *névé* which supplies it, it appeared to me that the Viescher Glacier ought to discharge as much ice as the Aletsch; but this is an error due to the extent of *névé* which is here at once visible: since a glance at the map of this portion of the Oberland shows at once the great superiority of the mountain treasury from which the Aletsch Glacier draws support. Still, the ice-field before us was a most noble one. The surrounding mountains were of imposing magnitude, and loaded to their summits with snow. Down the sides of some of them the half-consolidated mass fell in a state of wild fracture and confusion. In some cases the riven masses were twisted and overturned, the ledges bent, and the detached blocks piled one upon another in heaps; while in other cases the smooth white mass descended from crown to base without a wrinkle. The valley now below us was gorged by the frozen material thus incessantly poured into it. We crossed it, and reached the base of the Finsteraarhorn, ascended the mountain a little way, and at six o'clock paused to lighten our burdens and to refresh ourselves.

The north wind had freshened, we were in the shade, and the cold was very keen. Placing a bottle of tea and a small quantity of provisions in the knapsack, and a few figs and dried prunes in our pockets, we commenced the ascent. The Finsteraarhorn sends down a number of cliffy buttresses, separated from each other by wide couloirs filled with ice and snow. We ascended one of these buttresses for a time, treading cautiously among the spiky rocks; afterwards we went along the snow at the edge of the spine, and then fairly parted company with the rock, abandoning ourselves to the *névé* of the couloir. The latter was steep, and the snow was so firm that steps had to be cut in it. Once I paused upon a little ledge, which gave me a slight footing, and took the inclination. The slope formed an angle of 45° with the horizon; and across it, at a little distance below me, a gloomy fissure opened its jaws. The sun now cleared the summits which had before cut off his rays, and burst upon us with great power, compelling us to resort to our veils and dark spectacles. Two years before, Bennen had been nearly blinded by inflammation brought on by the glare from the snow, and he now took unusual care in protecting his eyes. The rocks looking more practicable, we again made towards them, and clambered among them till a vertical precipice, which proved impossible of ascent, fronted us. Bennen scanned the obstacle closely as we slowly approached it, and finally descended to the snow, which wound at a steep angle round its base: on this the footing appeared to me to be singularly insecure, but I marched without hesitation or anxiety in the footsteps of my guide.

We ascended the rocks once more, continued along them for some time, and then deviated to the couloir on our left. This snow-slope is much dislocated at its lower portion, and above its precipices and crevasses our route now lay. The snow was smooth, and sufficiently firm and steep to render the cutting of steps necessary. Bennen took the lead: to make each step he swung his mattock once, and his hindmost foot rose exactly at the moment the mattock descended; there was thus a kind of rhythm in his motion, the raising of the foot keeping time to the swing of the implement. In this manner we proceeded till we reached the base of the rocky pyramid which caps the mountain.

One side of the pyramid had been sliced off, thus dropping down almost a sheer precipice for some thousands of feet to the Finsteraar Glacier. A wall of rock, about 10 or 15 feet high, runs along the edge of the mountain, and this sheltered us from the north wind, which surged with the sound of waves against the tremendous barrier at the other side. "Our hardest work is now before us," said my guide. Our way lay up the steep and splintered rocks, among which we sought out the spikes which were closely enough wedged to bear our weight. Each had to trust to himself, and I fulfilled to the letter my engagement with Bennen to ask no help. My boiling-water apparatus and telescope were on my back, much to my annoyance, as the former was heavy, and sometimes swung awkwardly round as I twisted myself among the cliffs. Bennen offered to take it, but he had his own share to carry, and I was resolved to bear mine. Sometimes the rocks alternated with spaces of ice and snow, which we were at intervals compelled to cross; sometimes, when the slope was pure ice and very steep, we were compelled to retreat to the highest cliffs. The wall to which I have referred had given way in some places, and through the gaps thus formed the wind rushed with a loud, wild, wailing sound. Through these spaces I could see the entire field of Agassiz's observations; the junction of the Lauteraar and Finsteraar Glaciers at the Abschwung, the medial moraine between them, on which stood the Hôtel des Neufchâtelois, and the pavilion built by M. Dolfuss, in which Huxley and myself had found shelter two years before. Bennen was evidently anxious to reach the summit, and recommended all observations to be postponed until after our success had been assured. I agreed to this, and kept close at his heels. Strong as he was, he sometimes paused, laid his head upon his mattock, and panted like a chased deer. He complained of fearful thirst, and to quench it we had only my bottle of tea: this we shared loyally, my guide praising its virtues, as well he might. Still the summit loomed above us; still the angry swell of the north wind, beating against the torn battlements of the mountain, made wild music. Upward, however, we strained; and at last, on gaining the crest of a rock, Bennen exclaimed, in a jubilant voice, "*Die höchste Spitze!*"—the highest point. In a moment I was at his side, and saw the summit within a few paces of us. A minute or two placed us upon the topmost pinnacle, with the blue dome of heaven above us, and a world of mountains, clouds, and glaciers beneath.

A notion is entertained by many of the guides that if you go to sleep at the summit of any of the highest mountains, you will

"Sleep the sleep that knows no waking."

Bennen did not appear to entertain this superstition; and before starting in the morning, I had stipulated for ten minutes' sleep on reaching the summit, as part compensation for the loss of the night's rest. My first act, after casting a glance over the glorious scene beneath us, was to take advantage of this agreement; so I lay down and had five minutes' sleep, from which I rose refreshed and brisk. The sun at first beat down upon us with intense force, and I exposed my thermometers; but thin veils of vapour soon drew themselves before the sun, and denser mists spread over the valley of the Rhone, thus destroying all possibility of concert between Ramsay and myself. I turned therefore to my boiling-water apparatus, filled it with snow, melted the first charge, put more in, and boiled it; ascertaining the boiling point to be 187° Fahrenheit. On a sheltered ledge, about two or three yards south of the highest point, I placed a minimum-thermometer, in the hope that it would enable us in future years to record the lowest winter temperatures at the summit of the mountain.[1]

[1] The following note describes the single observation made with this thermometer. Mr. B. informs me that on finding the instrument Bennen swung it in triumph round his head. I fear, therefore, that the observation gives us no certain information regarding the minimum winter temperature.

"ST. NICHOLAS, 1859, *Aug.* 25.

"SIR,—On Tuesday last (the 23rd inst.) a party, consisting of Messrs. B., H., R. L., and myself, succeeded in reaching the summit of the Finster Aarhorn under the guidance of Bennen and Melchior André. We made it an especial object to observe and reset the minimum-thermometer which you left there last year. On reaching the summit, before I had time to stop him, Bennen produced the instrument, and it is just possible that in moving it he may have altered the position of the index. However, as he held the instrument horizontally, and did not, as far as I saw, give it any sensible jerk, I have great confidence that the index remained unmoved.

"The reading of the index was − 32° Cent.

"A portion of the spirit extending over about $10\frac{1}{2}°$ (and standing between 33° and $43\frac{1}{2}°$) was separated from the rest, but there appeared to be no data for determining when the separation had taken place. As it appeared desirable to unite the two portions of spirit before again setting the index to record the cold of another winter, we endeavoured to effect this by heating the bulb, but unfortunately, just as we were expecting to see them coalesce, the bulb burst, and I have now to express my great regret that my clumsiness or ignorance of the proper mode of setting the instrument in order should have interfered with the continuance of observations of so much interest. The remains of the instrument, together with a note of the accident, I have left in the charge of Wellig, the landlord of the hotel on the Æggischhorn.

"We reached the summit about 10.40 A.M. and remained there till noon; the reading of a pocket thermometer in the shade was 41° F.

"Should there be any further details connected with our ascent on which you would like to have information, I shall be happy to

It is difficult to convey any just impression of the scene from the summit of the Finsteraarhorn: one might, it is true, arrange the visible mountains in a list, stating their heights and distances, and leaving the imagination to furnish them with peaks and pinnacles, to build the precipices, polish the snow, rend the glaciers, and cap the highest summits with appropriate clouds. But if imagination did its best in this way, it would hardly exceed the reality, and would certainly omit many details which contribute to the grandeur of the scene itself. The various shapes of the mountains, some grand, some beautiful, bathed in yellow sunshine, or lying black and riven under the frown of impervious cumuli; the pure white peaks, cornices, bosses, and amphitheatres; the blue ice rifts, the stratified snow-precipices, the glaciers issuing from the hollows of the eternal hills, and stretching like frozen serpents through the sinuous valleys; the lower cloud field—itself an empire of vaporous hills—shining with dazzling whiteness, while here and there grim summits, brown by nature, and black by contrast, pierce through it like volcanic islands through a shining sea,—add to this the consciousness of one's position which clings to one *unconsciously*, that undercurrent of emotion which surrounds the question of one's personal safety, at a height of more than 14,000 feet above the sea, and which is increased by the weird strange sound of the wind surging with the full deep boom of the distant sea against the precipice behind, or rising to higher cadences as it forces itself through the crannies of the weatherworn rocks,—all conspire to render the scene from the Finsteraarhorn worthy of the monarch of the Bernese Alps.

My guide at length warned me that we must be moving; repeating the warning more impressively before I attended to it. We packed up, and as we stood beside each other ready to march he asked me whether we

supply them to the best of my recollection. Meanwhile, with a farther apology for my clumsiness, I beg to subscribe myself yours respectfully, H.

"Professor Tyndall."

Ascent of the Finsteraarhorn, 1858

should tie ourselves together, at the same time expressing his belief that it was unnecessary. Up to this time we had been separate, and the thought of attaching ourselves had not occurred to me till he mentioned it. I thought it, however, prudent to accept the suggestion, and so we united our destinies by a strong rope. "Now," said Bennen, "have no fear; no matter how you throw yourself, I will hold you." Afterwards, on another perilous summit, I repeated this saying of Bennen's to a strong and active guide, but his observation was that it was a hardy untruth, for that in many places Bennen could not have held me. Nevertheless a daring word strengthens the heart, and, though I felt no trace of that sentiment which Bennen exhorted me to banish, and was determined, as far as in me lay, to give him no opportunity of trying his strength in saving me, I liked the fearless utterance of the man, and sprang cheerily after him. Our descent was rapid, apparently reckless, amid loose spikes, boulders, and vertical prisms of rock, where a false step would assuredly have been attended with broken bones; but the consciousness of certainty in our movements never forsook us, and proved a source of keen enjoyment. The senses were all awake, the eye clear, the heart strong, the limbs steady, yet flexible, with power of recovery in store, and ready for instant action should the footing give way. Such is the discipline which a perilous ascent imposes.

We finally quitted the crest of rocks, and got fairly upon the snow once more. We first went downwards at a long swinging trot. The sun having melted the crust which we were compelled to cut through in the morning, the leg at each plunge sank deeply into the snow; but this sinking was partly in the direction of the slope of the mountain, and hence assisted our progress. Sometimes the crust was hard enough to enable us to glide upon it for long distances while standing erect; but the end of these *glissades* was always a plunge and tumble in the deeper snow. Once upon a steep hard slope Bennen's footing gave way; he fell, and went down rapidly, pulling me after him. I fell also, but turning quickly, drove the spike of my hatchet into the ice, got good anchorage, and held both fast; my success assuring me that I had improved as a mountaineer since my ascent of Mont Blanc. We tumbled so often in the soft snow, and our clothes and boots were so full of it, that we thought we might as well try the sitting posture in gliding down. We did so, and descended with extraordinary velocity, being checked at intervals by a bodily immersion in the softer and deeper snow. I was usually in front of Bennen, shooting down with the speed of an arrow, and feeling the check of the rope when the rapidity of my motion exceeded my guide's estimate of what was safe. Sometimes I was behind him, and darted at intervals with the swiftness of an avalanche right upon him; sometimes in the same transverse line with him, with the full length of the rope between us; and here I found its check unpleasant, as it tended to make me roll over. My feet were usually in the air, and it was only necessary to turn them right or left, like the helm of a boat, to change the direction of motion and avoid a difficulty, while a vigorous dig of leg and hatchet into the snow was sufficient to check the motion and bring us to rest. Swiftly, yet cautiously, we glided into the region of crevasses, where we at last rose, quite wet, and resumed our walking, until we reached the point where we had left our wine in the morning, and where I squeezed the water from my wet clothes, and partially dried them in the sun.

We had left some things at the cave of the Faulberg, and it was Bennen's first intention to return that way and take them home with him. Finding, however, that we could traverse the Viescher Glacier almost to the Æggischhorn, I made this our highway homewards. At the place where we entered it, and for an hour or two afterwards, the glacier was cut by fissures, for the most part covered with snow. We had packed up our rope, and Bennen admonished me to tread in his steps. Three or four times he half disappeared in the concealed fissures, but by clutching the snow he rescued himself and went on as swiftly as before. Once my leg sank, and the ring of icicles some fifty feet below told me that I was in the jaws of a crevasse; my guide turned sharply—it was the only time that I had seen concern on his countenance:—

"*Gott's Donner! Sie haben meine Tritte nicht gefolgt.*"

"*Doch!*" was my only reply, and we went on. He scarcely tried the snow that he crossed, as from its form and colour he could in most cases judge of its condition. For a long time we kept at the left-hand side of the glacier, avoiding the fissures which were now permanently open. We came upon the tracks of a herd of chamois, which had clambered from the glacier up the sides of the Oberaarhorn, and afterwards crossed the glacier to the right-hand side, my guide being perfect master of the ground. His eyes went in advance of his steps, and his judgment was formed before his legs moved. The glacier was deeply fissured, but there was no swerving, no retreating, no turning back to seek more practicable routes; each stride told, and every stroke of the axe was a profitable investment of labour.

We left the glacier for a time, and proceeded along the mountain side, till we came near the end of the Trift Glacier, where we let ourselves down an awkward face of rock along the track of a little cascade, and came upon the glacier once more. Here again I had occasion to admire the knowledge and promptness of my guide. The glacier, as is well known, is greatly dislocated, and has once or twice proved a prison to guides and travellers, but Bennen led me through the confusion without a pause. We were sometimes in the middle of the glacier, sometimes on the moraine, and sometimes on the side of the flanking mountain. Towards the end of the day we crossed what seemed to be the consolidated remains of a great avalanche; on this my foot slipped, there was a crevasse at hand, and a sudden effort was necessary to save me from falling into it. In making this effort the spike of my axe turned uppermost, and the palm of my hand came down upon it, thus inflicting a very angry wound. We were soon upon the green alp, having bidden a last farewell to the ice. Another hour's hard walking brought us to our hotel. No one seeing us crossing the alp would have supposed that we had laid such a day's work behind us; the proximity of home gave vigour to our strides, and our progress was much more speedy than it had been on starting in the morning. I was affectionately welcomed by Ramsay, had a warm bath, dined, went to bed, where I lay fast locked in sleep for eight hours, and rose next morning as fresh and vigorous as if I had never scaled the Finsteraarhorn.

XVII

On the 6th of August there was a long fight between mist and sunshine, each triumphing by turns, till at length the orb gained the victory and cleansed the mountains from every trace of fog. We descended to the Märjelen See, and, wishing to try the floating power of its icebergs, at a place where masses sufficiently large approached near to the shore, I put aside a portion of my clothes, and retaining my boots, stepped upon the floating ice. It bore me for a time, and I hoped eventually to be able to paddle myself over the water. On swerving a little, however, from the position in which I first stood, the mass turned over and let me into the lake. I tried a second one, which served me in the same manner; the water was too cold to continue the attempt, and there was also some risk of being unpleasantly ground between the opposing surfaces of the masses of ice. A very large iceberg which had been detached some short time previously from the glacier lay floating at some distance from us. Suddenly a sound like that of a waterfall drew our attention towards it. We saw it roll over with the utmost deliberation, while the water which it carried along with it rushed in cataracts down its sides. Its previous surface was white, its present one was of a lovely blue, the submerged crystal having now come to the air. The summerset of this iceberg produced a commotion all over the lake; the floating masses at its edge clashed together, and a mellow glucking sound, due to the lapping of the undulations against the frozen masses, continued long afterwards.

We subsequently spent several hours upon the glacier; and on this day I noticed for the first time a contemporaneous exhibition of *bedding* and *structure* to which I shall refer at another place. We passed finally to the left bank of the glacier, at some distance below the base of the Æggischhorn, and traced its old moraines at intervals along the flanks of the bounding mountain. At the summit of the ridge we found several fine old *roches moutonnées,* on some of which the scratchings of a glacier long departed were well preserved; and from the direction of the scratchings it might be inferred that the ice moved down the mountain towards the valley of the Rhone. A plunge into a lonely mountain lake ended the day's excursion.

On the 7th of August we quitted this noble station. Sending our guide on to Viesch to take a conveyance and proceed with our luggage down the valley, Ramsay and myself crossed the mountains obliquely, desiring to trace the glacier to its termination. We had no path, but it was hardly possible to go astray. We crossed spurs, climbed and descended pleasant mounds, sometimes with the soft grass under our feet, and sometimes knee-deep in rhododendrons. It took us several hours to reach the end of the glacier, and we then looked down upon it merely. It lay couched like a reptile in a wild gorge, as if it had split the mountain by its frozen snout. We afterwards descended to Mörill, where we met our guide and driver; thence down the valley to Visp; and the following evening saw us lodged at the Monte Rosa Hotel in Zermatt.

The boiling-point of water on the table of the *salle à manger* I found to be 202.58° Fahr.

On the following morning I proceeded without my friend to the Görner Glacier. As is well known, the end of this glacier has been steadily advancing for several years, and when I saw it, the meadow in front of it was partly shrivelled up by its irresistible advance. I was informed by my host that within the last sixty years forty-four châlets had been overturned by the glacier, the ground on which they stood being occupied by the ice; at present there are others for which a similar fate seems imminent. In thus advancing the glacier merely takes up ground which belonged to it in former ages, for the rounded rocks which rise out of the adjacent meadow show that it had once passed over them.

I had arranged to meet Ramsay this morning on the road to the Riffelberg. The meeting took place, but I then learned that a minute or two after my departure he had received intelligence of the death of a near relative. Thus was our joint expedition terminated, for he resolved to return at once to England. At my solicitation he accompanied me to the Riffel Hotel. We had planned an ascent of Monte Rosa together, but the arrangement thus broke down, and I was consequently thrown upon my own resources. Lauener had never made the ascent, but he nevertheless felt confident that we should accomplish it together.

FIRST ASCENT OF MONTE ROSA, 1858

XVIII

On Monday, the 9th of August, we reached the Riffel, and, by good fortune, on the evening of the same day, my guide's brother, the well-known Ulrich Lauener, also arrived at the hotel on his return from Monte Rosa. From him we obtained all the information possible respecting the ascent, and he kindly agreed to accompany us a little way the next morning, to put us on the right track. At three A.M. the door of my bedroom opened, and Christian Lauener announced to me that the weather was sufficiently good to justify an attempt. The stars were shining overhead; but Ulrich afterwards drew our attention to some heavy clouds which clung to the mountains on the other side of the valley of the Visp; remarking that the weather *might* continue fair throughout the

First Ascent of Monte Rosa, 1858

day, but that these clouds were ominous. At four o'clock we were on our way, by which time a grey stratus cloud had drawn itself across the neck of the Matterhorn, and soon afterwards another of the same nature encircled his waist. We proceeded past the Riffelhorn to the ridge above the Görner Glacier, from which Monte Rosa was visible from top to bottom, and where an animated conversation in Swiss patois commenced. Ulrich described the slopes, passes, and precipices, which were to guide us; and Christian demanded explanations, until he was finally able to declare to me that his knowledge was sufficient. We then bade Ulrich good-bye, and went forward. All was clear about Monte Rosa, and the yellow morning light shone brightly upon its uppermost snows. Beside the Queen of the Alps was the huge mass of the Lyskamm, with a saddle stretching from the one to the other; next to the Lyskamm came two white rounded mounds, smooth and pure, the Twins Castor and Pollux, and further to the right again the broad brown flank of the Breithorn. Behind us Mont Cervin gathered the clouds more thickly round him, until finally his grand obelisk was totally hidden. We went along the mountain side for a time, and then descended to the glacier. The surface was hard frozen, and the ice crunched loudly under our feet. There was a hollowness and volume in the sound which require explanation; and this, I think, is furnished by the remarks of Sir John Herschel on those hollow sounds at the Solfaterra, near Naples, from which travellers have inferred the existence of cavities within the mountain. At the place where these sounds are heard the earth is friable, and, when struck, the concussion is reinforced and lengthened by the partial echoes from the surfaces of the fragments. The conditions for a similar effect exist upon the glacier, for the ice is disintegrated to a certain depth, and from the innumerable places of rupture little reverberations are sent, which give a length and hollowness to the sound produced by the crushing of the fragments on the surface.

We looked to the sky at intervals, and once a meteor slid across it, leaving a train of sparks behind. The blue firmament, from which the stars shone down so brightly when we rose, was more and more invaded by clouds, which advanced upon us from our rear, while before us the solemn heights of Monte Rosa were bathed in rich yellow sunlight. As the day advanced the radiance crept down towards the valleys; but still those stealthy clouds advanced like a besieging army, taking deliberate possession of the summits, one after the other, while grey skirmishers moved through the air above us. The play of light and shadow upon Monte Rosa was at times beautiful, bars of gloom and zones of glory shifting and alternating from top to bottom of the mountain.

At five o'clock a grey cloud alighted on the shoulder of the Lyskamm, which had hitherto been warmed by the lovely yellow light. Soon afterwards we reached the foot of Monte Rosa, and passed from the glacier to a slope of rocks, whose rounded forms and furrowed surfaces showed that the ice of former ages had moved over them; the granite was now coated with lichens, and between the bosses where mould could rest were patches of tender moss. As we ascended a peal to the right announced the descent of an avalanche from the Twins; it came heralded by clouds of ice-dust, which resembled the sphered masses of condensed vapour which issue from a locomotive. A gentle snow-slope brought us to the base of a precipice of brown rocks, round which we wound; the snow was in excellent order, and the chasms were so firmly bridged by the frozen mass that no caution was necessary in crossing them. Surmounting a weathered cliff to our left, we paused upon the summit to look upon the scene around us. The snow gliding insensibly from the mountains, or discharged in avalanches from the precipices which it overhung, filled the higher valleys with pure white glaciers, which were rifted and broken here and there, exposing chasms and precipices from which gleamed the delicate blue of the half-formed ice. Sometimes, however, the *névés* spread over wide spaces without a rupture or wrinkle to break the smoothness of the superficial snow. The sky was now for the most part overcast, but through the residual blue spaces the sun at intervals poured light over the rounded bosses of the mountain.

At half-past seven o'clock we reached another precipice of rock, to the left of which our route lay, and here Lauener proposed to have some refreshment; after which we went on again. The clouds spread more and more, leaving at length mere specks and patches of blue between them. Passing some high peaks, formed by the dislocation of the ice, we came to a place where the *névé* was rent by crevasses on the walls of which the stratification due to successive snowfalls was shown with great beauty and definition. Between two of these fissures our way now lay: the wall of one of them was hollowed out longitudinally midway down, thus forming a roof above and a ledge below, and from roof to ledge stretched a railing of cylindrical icicles, as if intended to bolt them together. A cloud now for the first time touched the summit of Monte Rosa, and sought to cling to it, but in a minute it dispersed in shattered fragments, as if dashed to pieces for its presumption. The mountain remained for a time clear and triumphant, but the triumph was short-lived: like suitors that will not be repelled, the dusky vapours came; repulse after repulse took place, and the sunlight gushed down upon the heights, but it was manifest that the clouds gained ground in the conflict.

Until about a quarter past nine o'clock our work was mere child's play, a pleasant morning stroll along the flanks of the mountain; but steeper slopes now rose above us, which called for more energy, and more care in the fixing of the feet. Looked at from below, some of these slopes appeared precipitous; but we were too well acquainted with the effect of fore-shortening to let this daunt us. At each step we dug our batons into the deep snow. When first driven in, the batons[1] *dipped* from us, but were brought, as we walked forward, to the vertical, and finally beyond it at the other side. The snow was thus forced aside, a rubbing of the staff against it,

---

[1] My staff was always the handle of an axe an inch or two longer than an ordinary walking-stick.

and of the snow-particles against each other, being the consequence. We had thus perpetual rupture and regelation; while the little sounds consequent upon rupture, reinforced by the partial echoes from the surfaces of the granules, were blended together to a note resembling the lowing of cows. Hitherto I had paused at intervals to make notes, or to take an angle; but these operations now ceased, not from want of time, but from pure dislike; for when the eye has to act the part of a sentinel who feels that at any moment the enemy may be upon him; when the body must be balanced with precision, and legs and arms, besides performing actual labour, must be kept in readiness for possible contingencies; above all, when you feel that your safety depends upon yourself alone, and that, if your footing gives way, there is no strong arm behind ready to be thrown between you and destruction; under such circumstances the relish for writing ceases, and you are willing to hand over your impressions to the safe keeping of memory.

From the vast boss which constitutes the lower portion of Monte Rosa cliffy edges run upwards to the summit. Were the snow removed from these we should, I doubt not, see them as toothed or serrated crags, justifying the term "*kamm*," or "comb," applied to such edges by the Germans. Our way now lay along such a kamm, the cliffs of which had, however, caught the snow, and been completely covered by it, forming an edge like the ridge of a house-roof, which sloped steeply upwards. On the Lyskamm side of the edge there was no footing, and, if a human body fell over here, it would probably pass through a vertical space of some thousands of feet, falling or rolling, before coming to rest. On the other side the snow-slope was less steep, but excessively perilous-looking, and intersected by precipices of ice. Dense clouds now enveloped us, and made our position far uglier than if it had been fairly illuminated. The valley below us was one vast cauldron, filled with precipitated vapour, which came seething at times up the sides of the mountain. Sometimes this fog would partially clear away, and the light would gleam upwards from the dislocated glaciers.

First Ascent of Monte Rosa, 1858

My guide continually admonished me to make my footing sure, and to fix at each step my staff firmly in the consolidated snow. At one place, for a short steep ascent, the slope became hard ice, and our position a very ticklish one. We hewed our steps as we moved upwards, but were soon glad to deviate from the ice to a position scarcely less awkward. The wind had so acted upon the snow as to fold it over the edge of the kamm, thus causing it to form a kind of cornice, which overhung the precipice on the Lyskamm side of the mountain. This cornice now bore our weight: its snow had become somewhat firm, but it was yielding enough to permit the feet to sink in it a little way, and thus secure us at least against the danger of slipping. Here also at each step we drove our batons firmly into the snow, availing ourselves of whatever help they could render. Once, while thus securing my anchorage, the handle of my hatchet went right through the cornice on which we stood, and, on withdrawing it, I could see through the aperture into the cloud-crammed gulf below. We continued ascending until we reached a rock protruding from the snow, and here we halted for a few minutes. Lauener looked upwards through the fog. "According to all description," he observed, "this ought to be the last kamm of the mountain; but in this obscurity we can see nothing." Snow began to fall, and we recommenced our journey, quitting the rocks and climbing again along the edge. Another hour brought us to a crest of cliffs, at which, to our comfort, the kamm appeared to cease, and other climbing qualities were demanded of us.

On the Lyskamm side, as I have said, rescue would be out of the question, should the climber go over the edge. On the other side of the edge rescue seemed possible, though the slope, as stated already, was most dangerously steep. I now asked Lauener what he would have done, supposing my footing to have failed on the latter slope. He did not seem to like the question, but said that he should have considered well for a moment and then have sprung after me; but he exhorted me to drive all such thoughts away. I laughed at him, and this did more to set his mind at rest than any formal profession of courage could have done. We were now among rocks: we climbed cliffs and descended them, and advanced sometimes with our feet on narrow ledges, holding tightly on to other ledges by our fingers; sometimes, cautiously balanced, we moved along edges of rock with precipices on both sides. Once, in getting round a crag, Lauener shook a book from his pocket; it was arrested by a rock about sixty or eighty feet below us. He wished to regain it, but I offered to supply its place, if he thought the descent too dangerous. He said he would make the trial, and parted from me. I thought it useless to remain idle. A cleft was before me, through which I must pass; so, pressing my knees and back against its opposite sides, I gradually worked myself to the top. I descended the other face of the rock, and then, through a second ragged fissure, to the summit of another pinnacle. The highest point of the mountain was now at hand, separated from me merely by a short saddle, carved by weathering out the crest of the mountain. I could hear Lauener clattering after me, through the rocks behind. I dropped down upon the saddle, crossed it, climbed the opposite cliff, and "*die höchste Spitze*" of Monte Rosa was won.

Lauener joined me immediately, and we mutually congratulated each other on the success of the ascent. The residue of the bread and meat was produced, and a bottle of tea was also appealed to. Mixed with a little cognac, Lauener declared that he had never tasted anything like it. Snow fell thickly at intervals, and the obscurity was very great; occasionally this would lighten and permit the sun to shed a ghastly dilute light upon us through the gleaming vapour. I put my boiling-water apparatus in order, and fixed it in a corner behind a ledge; the shelter was, however, insufficient, so I placed my hat above the vessel. The boiling-point was 184.92° Fahr., the ledge on which the instrument stood being 5 feet below the highest point of the mountain.

The ascent from the Riffel Hotel occupied us about seven hours, nearly two of which were spent upon the kamm and crest. Neither of us felt in the least degree

First Ascent of Monte Rosa, 1858

fatigued; I, indeed, felt so fresh, that had another Monte Rosa been planted on the first, I should have continued the climb without hesitation, and with strong hopes of reaching the top. I experienced no trace of mountain sickness, lassitude, shortness of breath, heart-beat, or headache; nevertheless the summit of Monte Rosa is 15,284 feet high, being less than 500 feet lower than Mont Blanc. It is, I think, perfectly certain, that the rarefaction of the air at this height is not sufficient of itself to produce the symptoms referred to; physical exertion must be superadded.

After a few fitful efforts to dispel the gloom, the sun resigned the dominion to the dense fog and the descending snow, which now prevented our seeing more than 15 or 20 paces in any direction. The temperature of the crags at the summit, which had been shone upon by the unclouded sun during the earlier portion of the day, was 60° Fahr.; hence the snow melted instantly wherever it came in contact with the rock. But some of it fell upon my felt hat, which had been placed to shelter the boiling-water apparatus, and this presented the most remarkable and beautiful appearance. The fall of snow was in fact a shower of frozen flowers. All of them were six-leaved; some of the leaves threw out lateral ribs like ferns, some were rounded, others arrowy and serrated, some were close, others reticulated, but there was no deviation from the six-leaved type. Nature seemed determined to make us some compensation for the loss of all prospect, and thus showered down upon us those lovely blossoms of the frost; and had a spirit of the mountain inquired my choice, the view, or the frozen flowers, I should have hesitated before giving up that exquisite vegetation. It was wonderful to think of, as well as beautiful to behold. Let us imagine the eye gifted with a microscopic power sufficient to enable it to see the molecules which composed these starry crystals; to observe the solid nucleus formed and floating in the air; to see it drawing towards it its allied atoms, and these arranging themselves as if they moved to music, and ended by rendering that music concrete. Surely such an exhibition of power, such an apparent demonstration of a resident intelligence in what we are accustomed to call "brute matter," would appear perfectly miraculous. And yet the reality would, if we could see it, transcend the fancy. If the Houses of Parliament were built up by the forces resident in their own bricks and lithologic blocks, and without the aid of hodman or mason, there would be nothing intrinsically more wonderful in the process than in the molecular architecture which delighted us upon the summit of Monte Rosa.

Twice or thrice had my guide warned me that we must think of descending, for the snow continued to fall heavily, and the loss of our track would be attended with imminent peril. We therefore packed up, and clambered downward among the crags of the summit. We soon left these behind us, and as we stood once more upon the kamm, looking into the gloom beneath, an avalanche let loose from the side of an adjacent mountain shook the air with its thunder. We could not see it, could form no estimate of its distance, could only hear its roar, which, coming to us through the darkness, had an undefinable element of horror in it. Lauener remarked, "I never hear those things without a shudder; the memory of my brother comes back to me at the same time." His brother, who was the best climber in the Oberland, had been literally broken to fragments by an avalanche on the slopes of the Jungfrau.

We had been separate coming up, each having trusted to himself, but the descent was more perilous, because it is more difficult to fix the heel of the boot than the toe securely in the ice. Lauener was furnished with a rope, which he now tied round my waist, and forming a noose at the other end, he slipped it over his arm. This to me was a new mode of attachment. Hitherto my guides in dangerous places had tied the ropes round *their* waists also. Simond had done it on Mont Blanc, and Bennen on the Finsteraarhorn, proving thus their willingness to share my fate whatever that might be. But here Lauener had the power of sending me adrift at any moment, should his own life be imperilled. I told him that his mode of attachment was new to me, but he assured me

First Ascent of Monte Rosa, 1858

that it would give him more power in case of accident. I did not see this at the time; but neither did I insist on his attaching himself in the usual way. It could neither be called anger nor pride, but a warm flush ran through me as I remarked, that I should take good care not to test his power of holding me. I believe I wronged my guide by the supposition that he made the arrangement with reference to his own safety, for all I saw of him afterwards proved that he would at any time have risked his life to save mine. The flush, however, did me good, by displacing every trace of anxiety, and the rope, I confess, was also a source of some comfort to me. We descended the kamm, I going first. "Secure your footing before you move," was my guide's constant exhortation, "and make your staff firm at each step." We were sometimes quite close upon the rim of the kamm on the Lyskamm side, and we also followed the depressions which marked our track along the cornice. This I now tried intentionally, and drove the handle of my axe through it once or twice. At two places in descending we were upon the solid ice, and these were some of the steepest portions of the kamm. They were undoubtedly perilous, and the utmost caution was necessary in fixing the staff and securing the footing. These, however, once past, we felt that the chief danger was over. We reached the termination of the edge, and although the snow continued to fall heavily, and obscure everything, we knew that our progress afterwards was secure. There was pleasure in this feeling; it was an agreeable variation of that grim mental tension to which I had been previously wound up, but which in itself was by no means disagreeable.

I have already noticed the colour of the fresh snow upon the summit of the Stelvio Pass. Since I observed it there it has been my custom to pay some attention to this point at all great elevations. This morning, as I ascended Monte Rosa, I often examined the holes made in the snow by our batons, but the light which issued from them was scarcely perceptibly blue. Now, however, a deep layer of fresh snow overspread the mountain, and the effect was magnificent. Along the kamm I was continually surprised and delighted by the blue gleams which issued from the broken or perforated stratum of new snow; each hole made by the staff was filled with a light as pure, and nearly as deep, as that of the unclouded firmament. When we reached the bottom of the kamm, Lauener came to the front, and tramped before me. As his feet rose out of the snow, and shook the latter off in fragments, sudden and wonderful gleams of blue light flashed from them. Doubtless the blue of the sky has much to do with mountain colouring, but in the present instance not only was there no blue sky, but the air was so thick with fog and descending snow-flakes, that we could not see twenty yards in advance of us. A thick fog, which wrapped the mountain quite closely, now added its gloom to the obscurity caused by the falling snow. Before we reached the base of the mountain the fog became thin, and the sun shone through it. There was not a breath of air stirring, and, though we stood ankle-deep in snow, the heat surpassed anything of the kind I had ever felt: it was the dead suffocating warmth of the interior of an oven, which encompassed us on all sides, and from which there seemed no escape. Our own motion through the air, however, cooled us considerably. We found the snow-bridges softer than in the morning, and consequently needing more caution; but we encountered no real difficulty among them. Indeed it is amusing to observe the indifference with which a snow-roof is often broken through, and a traveller immersed to the waist in the jaws of a fissure. The effort at recovery is instantaneous; half instinctively hands and knees are driven into the snow, and rescue is immediate. Fair glacier work was now before us; after which we reached the opposite mountain-slope, which we ascended, and then went down the flank of the Riffelberg to our hotel. The excursion occupied us eleven and a half hours.

XIX

On the afternoon of the 11th I made an attempt alone to ascend the Riffelhorn, and attained a considerable height; but I attacked it from the wrong side, and the fading light forced me to retreat. I found some agreeable people at the hotel on my return. One clergyman especially, with a clear complexion, good digestion, and bad lungs—of free, hearty, and genial manner—made himself extremely pleasant to us all. He appeared to bubble over with enjoyment, and with him and others on the morning of the 13th I walked to the Görner Grat, as it lay on the way to my work. We had a glorious prospect from the summit: indeed the assemblage of mountains, snow, and ice, here within view is perhaps without a rival in the world.[1] I shouldered my axe, and saying "good-bye," moved away from my companions.

"Are you going?" exclaimed the clergyman. "Give me one grasp of your hand before we part."

This was the signal for a grasp all round; and the hearty human kindness which thus showed itself contributed that day to make my work pleasant to me.

We proceeded along the ridge of the Rothe Kumm to a point which commanded a fine view of the glacier. The ice had been over these heights in ages past, for, although lichens covered the surfaces of the old rocks, they did not disguise the grooves and scratchings. The surface of the glacier was now about a thousand feet below us, and this it was our desire to attain. To reach it we had to descend a succession of precipices, which in general were weathered and rugged, but here and there, where the rock was durable, were fluted and grooved. Once or twice indeed we had nothing to cling to but the little ridges thus formed. We had to squeeze ourselves through narrow fissures, and often to get round overhanging ledges, where our main trust was in our feet, but where these had only

[1] In 1858 Mr. E. W. Cooke made a pencil-sketch of this splendid panorama, which is the best and truest that I have yet seen.

ledges an inch or so in width to rest upon. These cases were to me the most unpleasant of all, for they compelled the arms to take a position which, if the footing gave way, would necessitate a *wrench*, for which I entertain considerable abhorrence. We came at length to a gorge by which the mountain is rent from top to bottom, and into which we endeavoured to descend. We worked along its rim for a time, but found its smooth faces too deep. We retreated; Lauener struck into another track, and while he tested it I sat down near some grass tufts, which flourished on one of the ledges, and found the temperature to be as follows:—

Temperature of rock	42° C.
Of air an inch above the rock . . .	32
Of air a foot from rock	22
Of grass	25

The first of these numbers does not fairly represent the temperature of the rock, as the thermometer could be in contact with it only at one side at a time. It was differences such as these between grass and stone, producing a mixed atmosphere of different densities, that weakened the sound of the falls of the Orinoco, as observed and explained by Humboldt.

By a process of "trial and error" we at length reached the ice, after two hours had been spent in the effort to disentangle ourselves from the crags. The glacier is forcibly thrust at this place against the projecting base of the mountain, and the structure of the ice correspondingly developed. Crevasses also intersect the ice, and the blue veins cross them at right angles. I ascended the glacier to a region where the ice was compressed and greatly contorted, and thought that in some cases I could see the veins crossing the lines of stratification. Once my guide drew my attention to what he called "*ein sonderbares Loch.*" On one of the slopes an archway was formed which appeared to lead into the body of the glacier. We entered it, and explored the cavern to its end. The walls were of transparent blue ice, singularly free from air-bubbles; but where the roof of the cavern was thin enough to allow the sun to shine feebly through it, the transmitted light was of a pink colour. My guide expressed himself surprised at "*das röthliche Schein.*" At one place a plate of ice had been placed like a ceiling across the cavern; but owing to lateral squeezing it had been broken so as to form a V. I found some air-bubbles in this ice, and in all cases they were associated with blebs of water. A portion of the "ceiling," indeed, was very full of bubbles, and was at some places reduced, by internal liquefaction, to a mere skeleton of ice, with water-cells between its walls.

High up the glacier (towards the old Weissthor) the horizontal stratification is everywhere beautifully shown. I drew my guide's attention to it, and he made the remark that the perfection of the lower ice was due to the pressure of the layers above it. "The snow by degrees compressed itself to glacier." As we approached one of the tributaries on the Monte Rosa side, where great pressure came into play, the stratification appeared to yield and the true structure to cross it at those places where it had yielded most. As the place of greatest pressure was approached, the bedding disappeared more and more, and a clear vertical structure was finally revealed.

THE GÖRNER GRAT AND THE RIFFEL-HORN. MAGNETIC PHENOMENA

XX

At an early hour on Saturday, the 14th of August, I heard the servant exclaim, "*Das Wetter ist wunderschön;*" which good news caused me to spring from my bed and prepare to meet the morn. The range of summits at the opposite side of the valley of St. Nicholas was at first quite clear, but as the sun ascended light cumuli formed round them, increasing in density up to a certain point; below these clouds the air of the valley was transparent; above them the air of heaven was still more so; and thus they swung midway between heaven and earth, ranging themselves in a level line along the necks of the mountains.

It might be supposed that the presence of the sun heating the air would tend to keep it more transparent, by increasing its capacity to dissolve all visible cloud; and this indeed is the true action of the sun. But it is not the only action. His rays, as he climbed the eastern heaven, shot more and more deeply into the valley of St. Nicholas, the moisture of which rose as invisible vapour, remaining unseen as long as the air possessed sufficient warmth to keep it in the vaporous state. High up, however, the cold crags which had lost their heat by radiation the night before, acted like condensers upon the ascending vapour, and caused it to curdle into visible fog. The current, however, continued ascensional, and the clouds were slowly lifted above the tallest peaks, where they arranged themselves in fantastic forms, shifting and changing shape as they gradually melted away. One peak stood like a field-officer with his cap raised above his head, others sent straggling cloud balloons upwards; but on watching these outliers they were gradually seen to disappear. At first they shone like snow in the sunlight, but as they became more attenuated they changed colour, passing through a dull red to a dusky purple hue, until finally they left no trace of their existence.

As the day advanced, warming the rocks, the clouds wholly disappeared, and a hyaline air formed the setting of both glaciers and mountains. I climbed to the Görner Grat to obtain a general view of the surrounding scene. Looking towards the origin of the Görner Glacier the view was bounded by a wide col, upon which stood two lovely rounded eminences enamelled with snow of perfect purity. They shone like burnished silver in the sunlight, as if their surfaces had been melted and recongealed to frosted mirrors from which the rays were flung. To the right of these were the bounding crags of Monte Rosa, and then the body of the mountain itself, with its crest of crag and coat of snows. To the right of Monte Rosa, and almost rivalling it in height, was the vast mass of the Lyskamm, a rough and craggy mountain, to whose ledges clings the snow which cannot grasp its steeper walls, sometimes leaning over them in impending precipices, which often break, and send wild avalanches into the space below. Between the Lyskamm and Monte Rosa lies a large wide valley into which both mountains pour their snows, forming there the Western Glacier of Monte Rosa—a noble ice stream, which from its magnitude and permanence deserves to impose its name upon the trunk glacier. It extends downwards from the col which unites the two mountains; riven and broken at some places, but at others stretching white and pure down to its snow-line, where the true glacier emerges from the *névé*. From the rounded shoulders of the Twin Castor a glacier descends, at first white and shining, then suddenly broken into faults, fissures, and precipices, which are afterwards repaired, and the glacier joins that of Monte Rosa before the junction of the latter with the trunk stream. Next came a boss of rock, with a secondary glacier clinging to it as if plastered over it, and after it the Schwarze Glacier, bounded on one side by the Breithorn, and on the other by the Twin Pollux. This glacier is of considerable magnitude. Over its upper portion rise the Twin eminences, pure and white; then follows a smooth and undulating space, after passing which the *névé* is torn up into a collection of peaks and chasms; these, however, are mended lower down, and the glacier moves smoothly and calmly to meet its brothers in the main valley. Next comes the Trifti Glacier,[1] embraced on all sides by the rocky arms of the Breithorn; its mass is not very great, but it descends in a graceful sweep, and exhibits towards its extremity a succession of beautiful bands. Afterwards we have the glacier of the Petit Mont Cervin and those of St. Theodule, which latter are the last that empty their frozen cargoes into the valley of the Görner. All the glaciers here mentioned are welded together to a common trunk which squeezes itself

---

[1] I take this name from Studer's map. Sometimes, however, I have called it the "Breithorn Glacier."

through the narrow defile at the base of the Riffelhorn. Soon afterwards the moraines become confused, the glacier drops steeply to its termination, and ploughs up the meadows in front of it with its irresistible share.

In a line with the Riffelhorn, and rising over the latter so high as to make it almost vanish by comparison, was the Titan obelisk of the Matterhorn, from the base of which the Furgge Glacier struggles downwards. On the other side are the Zmutt Glacier, the Schönbuhl, and the Hochwang, from the Dent Blanche; the Gabelhorn and Trift Glaciers, from the summits which bear those names. Then come the glaciers of the Weisshorn. Describing a curve still farther to the right we alight on the peaks of the Mischabel, dark and craggy precipices from this side, though from the Æggischhorn they appear as cones of snow. Sweeping by the Alphubel, the Allaleinhorn, the Rympfischorn, and Strahlhorn—all of them majestic—we reach the pass of the Weissthor, and the Cima di Jazzi. This completes the glorious circuit within the observer's view.

I placed my compass upon a piece of rock to find the bearing of the Görner Glacier, and was startled at seeing the sun and it at direct variance. What the sun declared to be north, the needle affirmed to be south. I at first supposed that the maker had placed the S where the N ought to be, and *vice versâ*. On shifting my position, however, the needle shifted also, and I saw immediately that the effect was due to the rock of the Grat. Sometimes one end of the needle *dipped* forcibly, at other places it whirled suddenly round, indicating an entire change of polarity. The rock was evidently to be regarded as an assemblage of magnets, or as a single magnet full of "consequent points." A distance of transport not exceeding an inch was, in some cases, sufficient to reverse the position of the needle. I held the needle between the two sides of a long fissure a foot wide. The needle set *along* the fissure at some places, while at others it set *across* it. Sometimes a little jutting knob would attract the north end of the needle, while a closely adjacent little knob would forcibly repel it, and attract the south end.

One extremity of a ledge three feet long was north magnetic, the other end was south magnetic, while a neutral point existed midway between both, the ledge having therefore the exact polar arrangement of an ordinary bar-magnet. At the highest point of the rock the action appeared to be most intense, but I also found an energetic polarity in a mass at some distance below the summit.

Remembering that Professor Forbes had noticed some peculiar magnetic effect upon the Riffelhorn, I resolved to ascend it. Descending from the Grat, we mounted the rocks which form the base of the horn; these are soft and soapy from the quantity of mica which they contain; the higher rocks of the horn are, however, very dense and hard. The ascent is a pleasant bit of mountain practice. We climbed the walls of rock, and wound round the ledges, seeking the assailable points. I tried the magnetic condition of the rocks as we ascended, and found it in general feeble. In other respects the Riffelhorn is a most remarkable mass. The ice of the Görner Glacier of former ages, which rose hundreds, perhaps thousands of feet above its present level, encountered the horn in its descent, and was split by the latter, a diversion of the ice along the sides of the peak being the consequence. Portions of the vertical walls of the horn are polished by this action as if they had come from the hands of a lapidary, and the scratchings are as sharp and definite as if drawn by points of steel. I never saw scratchings so perfectly preserved: the finest lines are as clear as the deepest, a consequence of the great density and durability of the rock. The latter evidently contains a good deal of iron, and its surface near the summit is of the rich brown red due to the peroxide of the metal. When we fairly got among the precipices we left our hatchets behind us, trusting subsequently to our hands and feet alone. Squeezing, creeping, clinging, and climbing, in due time we found ourselves upon the summit of the horn.

A pile of stones had been erected near the point where we gained the top. I examined the stones of this pile, and found them strongly polar. The surrounding rocks also showed a violent action, the needle oscillating quickly, and sometimes twirling swiftly round upon a slight change of position. The fragments of rock scattered about were also polar. Long ledges showed north magnetism for a considerable length, and again for an equal length south magnetism. Two parallel masses separated from each other by a fissure, showed the same magnetic distribution. While I was engaged at one end of the horn, Lauener wandered to the other, on which stood two or three *hommes de pierres.* He was about disturbing some of the stones, when a yell from me surprised him. In fact, the thought had occurred to me that the magnetism of the horn had been developed by lightning striking upon it, and my desire was to examine those points which were most exposed to the discharge of the atmospheric electricity; hence my shout to my guide to let the stones alone. I worked towards the other end of the horn, examining the rocks in my way. Two weathered prominences, which seemed very likely recipients of the lightning, acted violently upon the needle. I sometimes descended a little way, and found that among the rocks below the summit the action was greatly enfeebled. On reaching another very prominent point, I found its extremity all north polar, but at a little distance was a cluster of consequent points, among which the transport of a few inches was sufficient to turn the needle round and round.

The piles of stone at the Zermatt end of the horn did not seem so strongly polar as the pile at the other end, which was higher; still a strong polar action was manifested at many points of the surrounding rocks. Having completed the examination of the summit, I descended the horn, and examined its magnetic condition as I went along. It seemed to me that the jutting prominences always exhibited the strongest action. I do not indeed remember any case in which a strong action did not exhibit itself at the ends of the terraces which constitute the horn. In all cases, however, the rock acted as a number of magnets huddled confusedly together, and not as if its entire mass was endowed with magnetism of one kind.

On the evening of the same day I examined the lower

The Görner Grat

spur of the Riffelhorn. Amid its fissures and gullies one feels as if wandering through the ruins of a vast castle or fortification; the precipices are so like walls, and the scratching and polishing so like what might be done by the hands of man. I found evidences of strong polar action in some of the rocks low down. In the same continuous mass the action would sometimes exhibit itself over an area of small extent, while the remainder of the rock showed no appreciable action. Some of the boulders cast down from the summit exhibited a strong and varied polarity. Fig. 8 is a sketch of one of these; the barbed end of each arrow represents the north end of the needle, which assumed the various positions shown in the figure. Midway down the spur I lighted upon a transverse wall of rock, which formed in earlier ages the boundary of a lateral outlet of the Görner Glacier. It was red and hard, weathered rough at some places, and polished smooth at others.

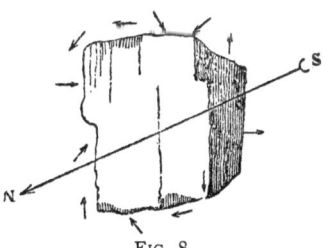

Fig. 8

The lines were drawn finely upon it, but its outer surface appeared to be peeling off like a crust; the polished layer rested upon the rock like a kind of enamel. The action of the glacier appeared to resemble that of the brake of a locomotive upon rails, both being cases of exfoliation brought about by pressure and friction. This wall measured twenty-eight yards across, and one end of it, for a distance of ten or twelve yards, was all north polar; the other end for a similar distance was south polar, but there was a pair of consequent points at its centre.

To meet the case of my young readers, I will here say a few words about the magnetic force. The common magnetic needle points nearly north and south; and if a bit of *iron* be brought near to either end of the needle, they will mutually attract each other. A piece of lead will not show this effect, nor will copper, gold, nor silver. Iron, in fact, is a *magnetic metal*, which the others are not. It is to be particularly observed, that the bit of iron attracts *both ends* of the needle when it is presented to them in succession; and if a common steel sewing needle be substituted for the iron it will be seen that it also has the power of attracting *both ends* of the magnetic needle. But if the needle be rubbed once or twice along one end of a magnet, it will be found that one of its ends will afterwards *repel* a certain end of the magnetic needle and attract the other. By rubbing the needle on the magnet, we thus develop both attraction and repulsion, and this double action of the magnetic force is called its *polarity*; thus the steel which was at first simply *magnetic*, is now magnetic and *polar*.

It is the aim of persons making magnets, that each magnet should have but *two* poles, at its two ends; but it is quite easy to develop in the same piece of steel several pairs of poles; and if the magnetisation be irregular, this is sometimes done when we wish to avoid it. These irregular poles are called *consequent points*.

Now I want my young reader to understand that it is not only because the rocks of the Görner Grat and Riffelhorn contain iron, that they exhibit the action which I have described. They are not only *magnetic*, as common iron is, but, like the magnetised steel needle, they are magnetic and polar. And these poles are irregularly distributed like the "consequent points" to which I have referred, and this is the reason why I have used the term.

Professor Forbes, as I have already stated, was the first to notice the effect of the Riffelhorn upon the magnetic needle, but he seems to have supposed that the entire mass of the mountain exercised "a local attraction" upon the needle (upon which end he does not say). To enable future observers to allow for this attraction, he took the bearing of several of the surrounding mountains from the Riffelhorn; but it is very probable that had he changed his position a few inches, and perfectly certain had he changed it a few yards, he would have found a set of bearings totally different from those which he has recorded. The close proximity and irregular distribution of its consequent points would prevent the Riffelhorn from exerting any appreciable influence on *a distant needle*, as in this case the local poles would effectually neutralise each other.

XXI

On the morning of the 15th the Riffelberg was swathed in a dense fog, through which heavy rain showered incessantly. Towards one o'clock the continuity of the grey mass was broken, and sky-gleams of the deepest blue were seen through its apertures; these would close up again, and others open elsewhere, as if the fog were fighting for existence with the sun behind it. The sun, however, triumphed, the mountains came more and more into view, and finally the entire air was swept clear. I went up to the Görner Grat in the afternoon, and examined more closely the magnetism of its rocks; here, as on the Riffelhorn, I found it most pronounced at the jutting prominences of the Grat. Can it be that the superior exposure is more favourable to the formation of the magnetic oxide of iron? I secured a number of fragments, which I still possess, and which act forcibly upon a magnetic needle. The sun was near the western horizon, and I remained alone upon the Grat to see his last beams illuminate the mountains, which, with one exception, was without a trace of cloud. This exception was the Matterhorn, the appearance of which was extremely instructive. The obelisk appeared to be divided in two halves by a vertical line drawn from its summit half way down, to the windward of which we had the bare cliffs of the mountain; and to the left of it a cloud which appeared to cling tenaciously to the rocks. In reality, however, there was no clinging; the condensed vapour incessantly got away, but it was ever renewed, and thus a river of cloud had been sent from the mountain over the valley of Aosta. The wind in fact blew lightly up the valley of St. Nicholas charged with moisture, and when the air that held it rubbed against the cold cone of the Matterhorn the vapour was chilled and precipitated in his lee. The summit seemed to smoke sometimes like a burning mountain; for immediately after its generation, the fog was drawn away in long filaments by the wind. As the sun sank lower the ruddiness of his light augmented, until these filaments resembled streamers of flame. The sun sank deeper, the light was gradually withdrawn, and where it had entirely vanished it left the mountain like a desolate old man whose

> "hoary hair
> Stream'd like a meteor in the troubled air."

For a moment after the sun had disappeared the scene was amazingly grand. The distant west was ruddy, copious grey smoke-wreaths were wafted from the mountains, while high overhead, in an atmospheric region which seemed perfectly motionless, floated a broad thin cloud, dyed with the richest iridescences. The colours were of the same character as those which I had seen upon the Aletschorn, being due to interference, and in point of splendour and variety far exceeded anything ever produced by the mere coloured light of the setting sun.

On the 16th I was early upon the glacier. It had frozen hard during the night, and the partially liberated streams flowed, in many cases, over their own ice. I took some clear plates from under the water, and found in them numerous liquid cells, each associated with an air-bubble or a vacuous spot. The most common shape of the cells was a regular hexagon, but there were all forms between the perfect hexagon and the perfect circle. Many cells had also crimped borders, intimating that their primitive form was that of a flower with six leaves. A plate taken from ice which was defended from the sunbeams by the shadow of a rock had no such cells: so that those that I observed were probably due to solar radiation.

My first aim was to examine the structure of the Görnerhorn Glacier, which descends the breast of Monte Rosa until it is abruptly cut off by the great Western

Glacier of the mountain. Between them is a moraine which is at once terminal as regards the former, and lateral as regards the latter. The ice is veined vertically along the moraine, the direction of the structure being parallel to the latter. I ascended the glacier, and found, as I retreated from the place where the thrust was most violent, that the structure became more feeble. From the glacier I passed to the rocks called *auf der Platte*, so as to obtain a general view of its terminal portion. The gradual perfecting of the structure as the region of pressure was approached was very manifest: the ice at the end seemed to wrinkle up in obedience to the pressure, the structural furrows, from being scarcely visible, became more and more decided, and the lamination underneath correspondingly pronounced, until it finally attained a state of great perfection.

I now quitted the rocks and walked straight across the Western Glacier of Monte Rosa to its centre, where I found the structure scarcely visible. I next faced the Görner Grat, and walked down the glacier towards the moraine which divides it from the Görner Glacier. The mechanical conditions of the ice here are quite evident; each step brought me to a place of greater pressure, and also to a place of more highly developed structure, until finally near to the moraine itself, and running parallel to it, a magnificent lamination was developed. Here the superficial groovings could be traced to great distances, and beside the moraine were boulders poised on pedestals of ice through which the blue veins ran. At some places the ice had been weathered into laminæ not more than a line in thickness.

I now recrossed the Monte Rosa Glacier to its junction with the Schwartze Glacier, which descends between the Twins and Breithorn. The structure of the Monte Rosa Glacier is here far less pronounced than at the other side, and the pressure which it endures is also manifestly less; the structure of the Schwartze Glacier is fairly developed, being here parallel to its moraine. The cliffs of the Breithorn are much exposed to weathering action, and boulders are copiously showered down upon the adjacent ice. Between the Schwartze Glacier and the glacier which descends from the breast of the Breithorn itself these blocks ride upon a spine of ice, and form a moraine of grand proportions. From it a fine view of the glacier is attainable, and the gradual development of its structure as the region of maximum pressure is approached is very plain. A number of gracefully curved undulations sweep across the Breithorn Glacier, which are squeezed more closely together as the moraine is approached. All the glaciers that descend from the flanking mountains of the Görner valley are suddenly turned aside where they meet the great trunk stream, and are reduced by the pressure to narrow stripes of ice separated from each other by parallel moraines.

I ascended the Breithorn Glacier to the base of an ice-fall, on one side of which I found large crumples produced by the pressure, the veined structure being developed at right angles to the direction of the latter. No such structure was visible above this place. The crumples were cut by fissures, perpendicular to which the blue veins ran. I now quitted the glacier, and clambered up the adjacent alp, from which a fine view of the general surface was attainable. As in the case of the Görnerhorn Glacier, the gradual perfecting of the structure was very manifest; the dirt, which first irregularly scattered over the surface, gradually assumed a striated appearance, and became more and more decided as the moraine was approached. I now descended from the alp, and endeavoured to measure some of the undulations ; proceeding afterwards to the junction of the Breithorn Glacier with that of St. Théodule. The end of the latter appears to be crumpled by its thrust against the former, and the moraine between them, instead of being raised, runs along a hollow which is flanked by the crumples on either side. The Breithorn Glacier became more and more attenuated, until finally it actually vanished under its own moraines. On the sides of the crevasses, by which the Théodule Glacier is here intersected, I thought I could plainly see two systems of veins cutting each other at an angle of fifteen or twenty degrees. Reaching the Görner Glacier, at a place where its dislocation was very great, I proceeded down it past the Riffelhorn, to a point where it seemed possible to scale the opposite mountain wall. Here I crossed the glacier, treading with the utmost caution along the combs of ice, and winding through the entanglement of crevasses until the spur of the Riffelhorn was reached ; this I climbed to its summit, and afterwards crossed the Green Alp to our hotel.

The foregoing good day's work was rewarded by a sound sleep at night. The tourists were called in succession next morning, but after each call I instantly subsided into deep slumber, and thus healthily spaced out the interval of darkness. Day at length dawned and gradually brightened. I looked at my watch and found it twenty minutes to six. My guide had been lent to a party of gentlemen who had started at three o'clock for the summit of Monte Rosa, and he had left with me a porter who undertook to conduct me to one of the adjacent glaciers. But as I looked from my window the unspeakable beauty of the morning filled me with a longing to see the world from the top of Monte Rosa. I was in exceedingly good condition—could I not reach the summit alone? Trained and indurated as I had been, I felt that the thing was possible ; at all events I could try, without attempting anything which was not clearly within my power.

SECOND ASCENT OF MONTE ROSA, 1858

XXII

Whether my exercise be mental or bodily, I am always most vigorous when cool. During my student life in Germany, the friends who visited me always complained of the low temperature of my room, and here among the Alps it was no uncommon thing for me to wander over the glaciers from morning till evening in my shirt-sleeves. My object now was to go as light as possible, and hence I left my coat and neckcloth behind me, trusting to the sun and my own motion to make good the calorific waste. After breakfast I poured what remained of my tea into a small glass bottle, an ordinary demi-bouteille, in fact; the waiter then provided me with a ham sandwich, and, with my scrip thus frugally furnished, I thought the heights of Monte Rosa might be won. I had neither brandy nor wine, but I knew the immense amount of mechanical force represented by four ounces of bread and ham, and I therefore feared no failure from lack of nutriment. Indeed, I am inclined to think that both guides and travellers often impair their vigour and render themselves cowardly and apathetic by the incessant "refreshing" which they deem it necessary to indulge in on such occasions.

The guide whom Lauener intended for me was at the door; I passed him and desired him to follow me. This he at first refused to do, as he did not recognise me in my shirt-sleeves; but his companions set him right, and he ran after me. I transferred my scrip to his shoulders, and led the way upward. Once or twice he insinuated that that was not the way to the Schwarze-See, and was probably perplexed by my inattention. From the summit of the ridge which bounds the Görner Glacier the whole grand panorama revealed itself, and on the higher slopes of Monte Rosa—so high, indeed, as to put all hope of overtaking them, or even coming near them, out of the question—a row of black dots revealed the company which had started at three o'clock from the hotel. They had made remarkably good use of their time, and I was afterwards informed that the cause of this was the intense cold, which compelled them to keep up the proper supply of heat by increased exertion. I descended swiftly to the glacier, and made for the base of Monte Rosa, my guide following at some distance behind me. One of the streams, produced by superficial melting, had cut for itself a deep wide channel in the ice; it was not too wide for a spring, and with the aid of a run I cleared it and went on. Some minutes afterwards I could hear the voice of my companion exclaiming, in a tone of expostulation, "No, no, I won't follow you there." He, however, made a circuit,

Second Ascent of Monte Rosa 135

and crossed the stream; I waited for him at the place where the Monte Rosa Glacier joins the rock, "*auf der Platte*," and helped him down the ice-slope. At the summit of these rocks I again waited for him. He approached me with some excitement of manner, and said that it now appeared plain to him that I intended to ascend Monte Rosa, but that he would not go with me. I asked him to accompany me to the summit of the next cliff, which he agreed to do; and I found him of some service to me. He discovered the faint traces of the party in advance, and, from his greater experience, could keep them better in view than I could. We lost them, however, near the base of the cliff at which we aimed, and I went on, choosing as nearly as I could remember the route followed by Lauener and myself a week previous, while my guide took another route, seeking for the traces. The glacier here is crevassed, and I was among the fissures some distance in advance of my companion. Fear was manifestly getting the better of him, and he finally stood still, exclaiming, "No man can pass there." At the same moment I discovered the trace, and drew his attention to it; he approached me submissively, said that I was quite right, and declared his willingness to go on. We climbed the cliff, and discovered the trace in the snow above it. Here I transferred the scrip and telescope to my own shoulders, and gave my companion a cheque for five francs. He returned, and I went on alone.

The sun and heaven were glorious, but the cold was nevertheless intense, for it had frozen bitterly the night before. The mountain seemed more noble and lovely than when I had last ascended it; and as I climbed the slopes, crossed the shining cols, and rounded the vast snow-bosses of the mountain, the sense of being alone lent a new interest to the glorious scene. I followed the track of those who preceded me, which was that pursued by Lauener and myself a week previously. Once I deviated from it to obtain a glimpse of Italy over the saddle which stretches from Monte Rosa to the Lyskamm. Deep below me was the valley, with its huge and dislocated *névé*, and the slope on which I hung was just sufficiently steep to keep the attention aroused without creating anxiety. I prefer such a slope to one on which the thought of danger cannot be entertained. I become more weary upon a dead level, or in walking up such a valley as that which stretches between Visp and Zermatt, than on a steep mountain side. The *sense* of weariness is often no index to the expenditure of muscular force: the muscles may be charged with force, and, if the nervous excitant be feeble, the strength lies dormant, and we are tired without exertion. But the thought of peril keeps the mind awake, and spurs the muscles into action; they move with alacrity and freedom, and the time passes swiftly and pleasantly.

Occupied with my own thoughts as I ascended, I sometimes unconsciously went too quickly, and felt the effects of the exertion. I then slackened my pace, allowing each limb an instant of repose as I drew it out of the snow, and found that in this way walking became rest. This is an illustration of the principle which runs throughout nature—to accomplish physical changes, *time* is necessary. Different positions of the limb require different molecular arrangements; and to pass from one to the other requires time. By lifting the leg slowly and allowing it to fall forward by its own gravity, a man may get on steadily for several hours, while a very slight addition to this pace may speedily exhaust him. Of course the normal pace differs in different persons, but in all the power of endurance may be vastly augmented by the prudent outlay of muscular force.

The sun had long shone down upon me with intense fervour, but I now noticed a strange modification of the light upon the slopes of snow. I looked upwards, and saw a most gorgeous exhibition of interference-colours. A light veil of clouds had drawn itself between me and the sun, and this was flooded with the most brilliant dyes. Orange, red, green, blue—all the hues produced by diffraction were exhibited in the utmost splendour. There seemed a tendency to form circular zones of colour round the sun, but the clouds were not sufficiently uniform to

Second Ascent of Monte Rosa 137

permit of this, and they were consequently broken into spaces, each steeped with the colour due to the condition of the cloud at the place. Three times during my ascent similar veils drew themselves across the sun, and at each passage the splendid phenomena were renewed. As I reached the middle of the mountain an avalanche was let loose from the sides of the Lyskamm; the thunder drew my eyes to the place; I saw the ice move, but it was only the tail of the avalanche; still the volume of sound told me that it was a huge one. Suddenly the front of it appeared from behind a projecting rock, hurling its ice-masses with fury into the valley, and tossing its rounded clouds of ice-dust high into the atmosphere. A wild long-drawn sound, multiplied by echoes, now descended from the heights above me. It struck me at first as a note of lamentation, and I thought that possibly one of the party which was now near the summit had gone over the precipice. On listening more attentively I found that the sound shaped itself into an English "Hurrah!" I was evidently nearing the party, and on looking upwards I could see them, but still at an immense height above me. The summit still rose before them, and I therefore thought the cheer premature. A precipice of ice was now in front of me, around which I wound to the right, and in a few minutes found myself fairly at the bottom of the Kamm.

I paused here for a moment, and reflected on the work before me. My head was clear, my muscles in perfect condition, and I felt just sufficient fear to render me careful. I faced the Kamm, and went up slowly but surely, and soon heard the cheer which announced the arrival of the party at the summit of the mountain. It was a wild, weird, intermittent sound, swelling or falling as the echoes reinforced or enfeebled it. In getting through the rocks which protrude from the snow at the base of the last spur of the mountain, I once had occasion to stoop my head, and, on suddenly raising it, my eyes swam as they rested on the unbroken slope of snow at my left. The sensation was akin to giddiness, but I believe it was chiefly due to the absence of any object upon the snow upon which I could converge the axes of my eyes. Up to this point I had eaten nothing. I now unloosed my scrip, and had two mouthfuls of sandwich and nearly the whole of the tea that remained. I found here that my load, light as it was, impeded me. When fine balancing is necessary, the presence of a very light load, to which one is unaccustomed, may introduce an element of danger, and for this reason I here left the residue of my tea and sandwich behind me. A long long edge was now in front of me, sloping steeply upwards. As I commenced the ascent of this, the foremost of those whose cheer had reached me from the summit some time previously, appeared upon the top of the edge, and the whole party was seen immediately afterwards dangling on the Kamm. We mutually approached each other. Peter Bohren, a well-known Oberland guide, came first, and after him came the gentleman in his immediate charge. Then came other guides with other gentlemen, and last of all my guide, Lauener, with his strong right arm round the youngest of the party. We met where a rock protruded through the snow. The cold smote my naked throat bitterly, so to protect it I borrowed a handkerchief from Lauener, bade my new acquaintances good-bye, and proceeded upwards. I was soon at the place where the snow-ridge joins the rocks which constitute the crest of the mountain; through these my way lay, every step I took augmenting my distance from all life, and increasing my sense of solitude. I went up and down the cliffs as before, round ledges, through fissures, along edges of rock, over the last deep and rugged indentation, and up the rocks at its opposite side, to the summit.

A world of clouds and mountains lay beneath me. Switzerland, with its pomp of summits, was clear and grand; Italy was also grand, but more than half obscured. Dark cumulus and dark crag vied in savagery, while at other places white snows and white clouds held equal rivalry. The scooped valleys of Monte Rosa itself were magnificent, all gleaming in the bright sunlight—tossed and torn at intervals, and sending from their rents and walls the magical blue of the ice. Ponderous *névés* lay upon the mountains, apparently motionless, but suggesting motion—sluggish, but indicating irresistible dynamic energy, which moved them slowly to their doom in the warmer valleys below. I thought of my position: it was the first time that a man had stood alone upon that wild peak, and were the imagination let loose amid the surrounding agencies, and permitted to dwell upon the perils which separated the climber from his kind, I dare say curious feelings might have been engendered. But I was prompt to quell all thoughts which might lessen my strength, or interfere with the calm application of it. Once indeed an accident made me shudder. While taking the cork from a bottle which is deposited on the top, and which contains the names of those who have ascended the mountain, my axe slipped out of my hand, and slid some thirty feet away from me. The thought of losing it made my flesh creep, for without it descent would be utterly impossible. I regained it, and looked upon it with an affection which might be bestowed upon a living thing, for it was literally my staff of life under the circumstances. One look more over the cloud-capped mountains of Italy, and I then turned my back upon them, and commenced the descent.

The brown crags seemed to look at me with a kind of friendly recognition, and, with a surer and firmer feeling than I possessed on ascending, I swung myself from crag to crag and from ledge to ledge with a velocity which surprised myself. I reached the summit of the Kamm, and saw the party which I had passed an hour and a half before, emerging from one of the hollows of the mountain; they had escaped from the edge which now lay between them and me. The thought of the possible loss of my axe at the summit was here forcibly revived, for without it I dared not take a single step. My first care was to anchor it firmly in the snow, so as to enable it to bear at times nearly the whole weight of my body. In some places, however, the anchor had but a loose hold; the "cornice" to which I have already referred became granular, and the handle of the axe went through it up to the head, still, however, remaining loose. Some amount of trust had thus to be withdrawn from the staff and placed in the limbs. A curious mixture of carelessness and anxiety sometimes fills the mind on such occasions. I often caught myself humming a verse of a frivolous song, but this was mechanical, and the substratum of a man's feelings under such circumstances is real earnestness. The precipice to my left was a continual preacher of caution, and the slope to my right was hardly less impressive. I looked down the former but rarely, and sometimes descended for a considerable time without looking beyond my own footsteps. The power of a thought was illustrated on one of these occasions. I had descended with extreme slowness and caution for some time, when looking over the edge of the cornice I saw a row of pointed rocks at some distance below me. These I felt must receive me if I slipped over, and I thought how before reaching them I might so break my fall as to arrive at them unkilled. This thought enabled me to double my speed, and as long as the spiky barrier ran parallel to my track I held my staff in one hand, and contented myself with a slight pressure upon it.

I came at length to a place where the edge was solid ice, which rose to the level of the cornice, the latter appearing as if merely stuck against it. A groove ran between the ice and snow, and along this groove I marched until the cornice became unsafe, and I had to betake myself to the ice. The place was really perilous, but, encouraging myself by the reflection that it would not last long, I carefully and deliberately hewed steps, causing them to dip a little inward, so as to afford a purchase for the heel of my boot, never forsaking one till the next was ready, and never wielding my hatchet until my balance was secured. I was soon at the bottom of the Kamm, fairly out of danger, and, full of glad vigour, I bore swiftly down upon the party in advance of me. It was an easy task to me to fuse myself amongst them as if I had been an old acquaintance, and we joyfully slid, galloped, and rolled together down the residue of the mountain.

The only exception was the young gentleman in

Lauener's care. A day or two previously he had, I believe, injured himself in crossing the Gemmi, and long before he reached the summit of Monte Rosa his knee swelled, and he walked with great difficulty. But he persisted in ascending, and Lauener, seeing his great courage, thought it a pity to leave him behind. I have stated that a portion of the Kamm was solid ice. On descending this, Mr. F.'s footing gave way, and he slipped forward. Lauener was forced to accompany him, for the place was too steep and slippery to permit of their motion being checked. Both were on the point of going over the Lyskamm side of the mountain, where they would have indubitably been dashed to pieces. "There was no escape there," said Lauener, in describing the incident to me subsequently, "but I saw a possible rescue at the other side, so I sprang to the right, forcibly swinging my companion round; but in doing so, the baton tripped me up; we both fell, and rolled rapidly over each other down the incline. I knew that some precipices were in advance of us, over which we should have gone, so, releasing myself from my companion, I threw myself in front of him, stopped myself with my axe, and thus placed a barrier before him." After some vain efforts at sliding down the slopes on a baton, in which practice I was fairly beaten by some of my new friends, I attached myself to the invalid, and walked with him and Lauener homewards. Had I gone forward with the foremost of the party, I should have completed the expedition to the summit and back in a little better than nine hours.

I think it right to say one earnest word in connection with this ascent; and the more so as I believe a notion is growing prevalent that half what is said and written about the dangers of the Alps is mere humbug. No doubt exaggeration is not rare, but I would emphatically warn my readers against acting upon the supposition that it is general. The dangers of Mont Blanc, Monte Rosa, and other mountains, are real, and, if not properly provided against, may be terrible. I have been much accustomed to be alone upon the glaciers, but sometimes, even when a guide was in front of me, I have felt an extreme longing to have a second one behind me. Less than two good ones I think an arduous climber ought not to have; and if climbing without guides were to become habitual, deplorable consequences would assuredly sooner or later ensue.

XXIII

The 18th of August I spent upon the Furgge Glacier at the base of Mont Cervin, and what it taught me shall be stated in another place. The evening of this day was signalised by the pleasant acquaintances which it gave me. It was my intention to cross the Weissthor on the morning of the 19th, but thunder, lightning, and heavy rain opposed the project, and with two friends I descended, amid pitiless rain, to Zermatt. Next day I walked by way of Stalden to Saas, where I made the acquaintance of Herr Imseng, the Curé, and on the 21st ascended to the Distel Alp. Near to this place the Allelein Glacier pushes its huge terminus right across the valley and dams up the streams descending from the mountains higher up, thus giving birth to a dismal lake. At one end of this stands the Mattmark Hotel, which was to be my headquarters for a few days.

I reached the place in good company. Near to the hotel are two magnificent boulders of green serpentine, which have been lodged there by one of the lateral glaciers; and two of the ladies desiring to ascend one of these rocks, a friend and myself helped them to the top. The thing was accomplished in a very spirited way. Indeed the general contrast, in regard to energy, between the maidens of the British Isles and those of the Continent and of America is extraordinary. Surely those who talk of this country being in its old age overlook the physical vigour of its sons and daughters. They are strong, but from a combination of the greatest forces we may obtain a small resultant, because the forces may act in opposite directions and partly neutralise each other. Herein, in fact, lies Britain's weakness; it is strength ill directed;

Second Ascent of Monte Rosa

and is indicative rather of the perversity of young blood than of the precision of mature years.

Immediately after this achievement I was forsaken by my friends, and remained the only occupant of the hotel. A dense grey cloud gradually filled the entire atmosphere, from which the rain at length began to gush in torrents. The scene from the windows of the hotel was of the most dismal character; the rain also came through the roof, and dripped from the ceiling to the floor. I endeavoured to make a fire, but the air would not let the smoke of the pine-logs ascend, and the biting of the hydrocarbons was excruciating to the eyes. On the whole, the cold was preferable to the smoke. During the night the rain changed to snow, and on the morning of the 22nd all the mountains were thickly covered. The grey delta through which a river of many arms ran into the Mattmark See was hidden; against some of the windows of the *salle à manger* the snow was also piled, obscuring more than half their light. I had sent my guide to Visp, and two women and myself were the only occupants of the place. It was extremely desolate—I felt, moreover, the chill of Monte Rosa in my throat, and the conditions were not favourable to the cure of a cold.

On the 23rd the Allelein Glacier was unfit for work; I therefore ascended to the summit of the Monte Moro, and found the Valais side of the pass in clear sunshine, while impenetrable fog met us on the Italian side. I examined the colour of the freshly fallen snow; it was not an ordinary blue, and was even more transparent than the blue of the firmament. When the snow was broken the light flashed forth; when the staff was dug into the snow and withdrawn, the blue gleam appeared; when the staff lay in a hole, although there might be a sufficient space all round it, the coloured light refused to show itself.

My cough kept me awake on the night of the 23rd, and my cold was worse next day. I went upon the Allelein Glacier, but found myself by no means so sure a climber as usual. The best guides find that their powers vary; they are not equally competent on all days. I have

144 Glaciers of the Alps

heard a celebrated Chamouni guide assert that a man's morale is different on different days. The morale in my case had a physical basis, and it probably has so in all. The Allelein Glacier, as I have said, crosses the valley and abuts against the opposite mountain; here it is forced to turn aside, and in consequence of the thrust and bending it is crumpled and crevassed. The wall of the Mattmark See is a fine glacier section: looked at from a distance, the ridges and fissures appear arranged like a fan. The structure of the crumpled ice varies from the vertical to the horizontal, and the ridges are sometimes split *along* the planes of structure. The aspect of this portion of the glacier from some of the adjacent heights is exceedingly interesting.

On the morning of the 25th I had two hours' clambering over the mountains before breakfast, and traced the action of ancient glaciers to a great height. The valley of Saas in this respect rivals that of Hasli; the flutings and polishings being on the grandest scale. After breakfast I went to the end of the Allelein Glacier, where the Saas Visp River rushes from it: the vault was exceedingly fine, being composed of concentric arches of clear blue ice. I spent several hours here examining the intimate structure of the ice, and found the vacuum disks which I shall describe at another place, of the greatest service to me. As at Rosenlaui and elsewhere, they here taught me that the glacier was composed of an aggregate of small fragments, each of which had a definite plane of crystallisation. Where the ice was partially weathered the surfaces of division between the fragments could be traced through the coherent mass, but on crossing these surfaces the direction of the vacuum disks changed, indicating a similar change of the planes of crystallisation. The blue veins of the glacier went through its component fragments irrespective of these planes. Sometimes the vacuum disks were parallel to the veins, sometimes across them, sometimes oblique to them.

Several fine masses of ice had fallen from the arch upon its floor, and these were disintegrated to the core. A kick, or a stroke of an axe, sufficed to shake masses almost a cubic yard in size into fragments varying not much on either side of a cubic inch. The veining was finely preserved on the concentric arches of the vault, and some of them apparently exhibited its abolition, or at least confusion, and fresh development by new conditions of pressure. The river being deep and turbulent this day, to reach its opposite side I had to climb the glacier and cross over the crown of its highest arch; this enabled me to get quite in front of the vault, to enter it, and closely inspect those portions where the structure appeared to change. I afterwards ascended the steep moraine which lies between the Allelein and the smaller glacier to the left of it, passing to the latter at intervals to examine its structure. I was at length stopped by the dislocated ice; and from the heights I could count a system of seven dirt bands, formed by the undulations on the surface of the glacier. On my return to the hotel I found there a number of well-known Alpine men who intended to cross the Adler Pass on the following day. Herr Imseng was there: he came to me full of enthusiasm, and asked me whether I would join him in an ascent of the Dom: we might immediately attack it, and he felt sure that we should succeed. The Dom is the highest of the Mischabel peaks, and is one of the grandest of the Alps. I agreed to join the Curé, and with this understanding we parted for the night.

Thursday, 26th August.—A wild stormy morning after a wild and rainy night: the Adler Pass being impassable, the mountaineers returned, and Imseng informed me that the Dom must be abandoned. He gave me the statistics of an avalanche which had fallen in the valley some years before. Within the memory of man Saas had never been touched by an avalanche, but a tradition existed that such a catastrophe had once occurred. On the 14th of March 1848, at eight o'clock in the morning, the Curé was in his room, when he heard the cracking of pine-branches, and inferred from the sound that an avalanche was descending upon the village. It dashed in the windows of his house and filled his rooms with snow; the sound it produced being sufficient to mask the crashing of the timbers of an adjacent house. Three persons were killed. On the 3rd of April 1849, heavy snow fell at Saas; the Curé waited until it had attained a depth of four feet, and then retreated to Fée. That night an avalanche descended, and in the line of its rush was a house in which five or six and twenty people had collected for safety: nineteen of them were killed. The Curé afterwards showed me the site of the house, and the direction of the avalanche. It passed through a pine wood; and on expressing my surprise that the trees did not arrest it, he replied that the snow was "quite like dust," and rushed among the trees like so much water. To return from Fée to Saas on the day following he found it necessary to carry two planks. Kneeling upon one of them, he pushed the other forward, and transferred his weight to it, drawing the other after him and repeating the same act. The snow was like flour, and would not otherwise bear his weight. Seeing no prospect of fine weather, I descended to Saas on the afternoon of the 26th. I was the only guest at the hotel; but during the evening I was gratified by the unexpected arrival of my friend Hirst, who was on his way over the Monte Moro to Italy.

For the last five days it had been a struggle between the north wind and the south, each edging the other by turns out of its atmospheric bed, and producing copious precipitation; but now the conflict was decided—the north had prevailed, and an almost unclouded heaven overspread the Alps. The few white fleecy masses that remained were good indications of the swift march of the wind in the upper air. My friend and I resolved to have at least one day's excursion together, and we chose for it the glacier of the Fée. Ascending the mountain by a well-beaten path, we passed a number of "Calvaries" filled with tattered saints and Virgins, and soon came upon the rim of a flattened bowl quite clasped by the mountains. In its centre was the little hamlet of Fée, round which were fresh green pastures, and beyond it the perpetual ice and snow. It was exceedingly picturesque —a scene of human beauty and industry where savagery alone was to be expected. The basin had been scooped by glaciers, and as we paused at its entrance the rounded and fluted rocks were beneath our feet. The Alphubel and the Mischabel raised their crowns to heaven in front of us; the newly fallen snow clung where it could to the precipitous crags of the Mischabel, but on the summits it was the sport of the wind. Sometimes it was borne straight upwards in long vertical striæ; sometimes the fibrous columns swayed to the right, sometimes to the left; sometimes the motion on one of the summits would quite subside; anon the white peak would appear suddenly to shake itself to dust, which it yielded freely to the wind. I could see the wafted snow gradually melt away, and again curdle up into true white cloud by precipitation; this in its turn would be pulled asunder like carded wool, and reduced a second time to transparent vapour.

In the middle of the ice of the Fée stands a green alp, not unlike the Jardin; up this we climbed, halting at intervals upon its grassy knolls to inspect the glacier. I aimed at those places where on *a priori* grounds I should have thought the production of the veined structure most likely, and reached at length the base of a wall of rock from the edge of which long spears of ice depended. Here my friend halted, while Lauener and myself climbed the precipice, and ascended to the summit of the alp. The snow was deep at many places, and our immersions in unseen holes very frequent. From the peak of the Fée Alp a most glorious view is obtained; in point of grandeur it will bear comparison with any in the Alps, and its seclusion gives it an inexpressible charm. We remained for half-an-hour upon the warm rock, and then descended. It was our habit to jump from the higher ledges into the deep snow below them, in which we wallowed as if it were flour; but on one of these occasions I lighted on a stone, and the shock produced a curious effect upon my hearing. I appeared suddenly to lose the power of appreciating deep sounds, while the shriller ones were comparatively unimpaired. After I rejoined my friend it required attention on my part to hear him when he spoke to me. This continued until I approached the end of the glacier, when suddenly the babblement of streams, and a world of sounds to which I had been before quite deaf burst in upon me. The deafness was probably due to a strain of the tympanum, such as we can produce artificially, and thus quench low sounds, while shrill ones are scarcely affected.

I was anxious to quit Saas early next morning, but the Curé expressed so strong a wish to show us what he called a *Schauderhaftes Loch*—a terrible hole—which he had himself discovered, that I consented to accompany him. We were joined by his assistant and the priest of Fée. The stream from the Fée Glacier has cut a deep channel through the rocks, and along the right-hand bank of the stream we ascended. It was very rough with fallen crags and fallen pines amid which we once or twice lost our way. At length we came to an aperture just sufficient to let a man's body through, and were informed by our conductor that our route lay along the little tunnel: he lay down upon his stomach and squeezed himself through it like a marmot. I followed him; a second tunnel, in which, however, we could stand upright, led into a spacious cavern, formed by the falling together of immense slabs of rock which abutted against each other so as to form a roof. It was the very type of a robber den; and when I remarked this, it was at once proposed to sing a verse from Schiller's play. The young clergyman had a powerful voice—he led and we all chimed in.

> "Ein frohes Leben führen wir,
> Ein Leben voller Wonne.
> Der Wald ist unser Nachtquartier,
> Bei Sturm und Wind hanthieren wir,
> Der Mond ist unsre Sonne."

Herr Imseng wore his black coat; the others had taken theirs off, but they wore their clerical hats, black breeches and stockings. We formed a singular group in a singular place, and the echoed voices mingled strangely with the gusts of the wind and the rush of the river.

Soon after I parted from my friend, and descended the valley to Visp, where I also parted with my guide. He had been with me from the 22nd of July to the 29th of August, and did his duty entirely to my satisfaction. He is an excellent iceman, and is well acquainted both with the glaciers of the Oberland and of the Valais. He is strong and good-humoured, and were I to make another expedition of the kind I don't think that I should take any guide in the Oberland in preference to Christian Lauener.

XXIV

It is a singular fact that as yet we know absolutely nothing of the winter temperature of any one of the high Alpine summits. No doubt it is a sufficient justification of our Alpine men, as regards their climbing, *that they like it.* This plain reason is enough; and no man who ever ascended that "bad eminence" Primrose Hill, or climbed to Hampstead Heath for the sake of a freer horizon, can consistently ask a better. As regards physical science, however, the contributions of our mountaineers have as yet been *nil*, and hence, when we hear of the scientific value of their doings, it is simply amusing to the climbers themselves. I do not fear that I shall offend them in the least by my frankness in stating this. Their pleasure is that of overcoming acknowledged difficulties, and of witnessing natural grandeur. But I would venture to urge that our Alpine men will not find their pleasure lessened by embracing a scientific object in their doings. They have the strength, the intelligence, and let them add to these the accuracy which physical science now demands, and they may contribute work of enduring value. Mr. Casella will gladly teach them the use of his minimum-thermometers; and I trust that the next seven years will not pass without making us acquainted with the winter temperature of every mountain of note in Switzerland.[1]

[1] I find with pleasure that my friend Mr. John Ball is now exerting himself in this direction.

I had thought of this subject since I first read the conjectures of De Saussure on the temperature of Mont Blanc; but in 1857 I met Auguste Balmat at the Jardin, and there learned from him that he entertained the idea of placing a self-registering thermometer at the summit of the mountain. Balmat was personally a stranger to me at the time, but Professor Forbes's writings had inspired me with a respect for him, which this unprompted idea of his augmented. He had procured a thermometer, the graduation of which, however, he feared was not low enough. As an encouragement to Balmat, and with the view of making his laudable intentions known, I communicated them to the Royal Society, and obtained from the Council a small grant of money to purchase thermometers and to assist in the expenses of an ascent. I had now the thermometers in my possession; and having completed my work at Zermatt and Saas, my next desire was to reach Chamouni and place the instruments on the top of Mont Blanc. I accordingly descended the valley of the Rhone to Martigny, crossed the Tête Noire, and arrived at Chamouni on the 29th of August 1858.

Balmat was engaged at this time as the guide of Mr. Alfred Wills, who, however, kindly offered to place him at my disposal; and also expressed a desire to accompany me himself and assist me in my observations. I gladly accepted a proposal which gave me for companion so determined a climber and so estimable a man. But Chamouni was rife with difficulties. In 1857 the Guide Chef had the good sense to give me considerable liberty of action. Now his mood was entirely changed: he had been "molested" for giving me so much freedom. I wished to have a boy to carry a small instrument for me up the Mer de Glace—he would not allow it; I must take a guide. If I ascended Mont Blanc he declared that I must take four guides; that, in short, I must in all respects conform to the rules made for ordinary tourists. I endeavoured to explain to him the advantages which Chamouni had derived from the labours of men of science; it was such men who had discovered it when it was unknown, and it was by their writings that the attention of the general public had been called towards it. It was a bad recompense, I urged, to treat a man of science as he was treating me. This was urged in vain: he shrugged his shoulders, was very sorry, but the thing could not be changed. I then requested to know his superior, that I might apply to him; he informed me that there were a President and Commission of Guides at Chamouni, who were the proper persons to decide the question, and he proposed to call them together on the 31st of August, at 7 P.M., on condition that I was to be present to state my own case. To this I agreed.

I spent that day quite alone upon the Mer de Glace, and climbed amid a heavy snow-storm to the Cleft station over Trélaporte. When I reached the Montanvert I was wet and weary, and would have spent the night there were it not for my engagement with the Guide Chef. I descended amid the rain, and at the appointed hour went to his bureau. He met me with a polite sympathetic shrug; explained to me that he had spoken to the Commission, but that it could not assemble *pour une chose comme ça;* that the rules were fixed, and I must abide by them. "Well," I responded, "you think you have done your duty; it is now my turn to perform mine. If no other means are available I will have this transaction communicated to the Sardinian Government, and I don't think that it will ratify what you have done." The Guide Chef evidently did not believe a word of it.

Previous to taking any further step I thought it right to see the President of the Commission of Guides, who was also Syndic of the commune. I called upon him on the morning of the 1st of September, and, assuming that he knew all about the transaction, spoke to him accordingly. He listened to me for a time, but did not seem to understand me, which I ascribed partly to my defective French pronunciation. I expressed a hope that he did comprehend me; he said he understood my words very well, but did not know their purport. In fact he had not heard a single word about me or my request. He stated with some indignation that, so far from its being a subject on which the Commission could not assemble, it was one which it was their especial duty to take into consideration. Our conference ended with the arrangement that I was to write him an official letter stating the case, which he was to forward to the Intendant of the province of Faucigny resident at Bonneville. All this was done.

I subsequently memorialised the Intendant himself; and Balmat visited him to secure his permission to accompany me. I have to record, that from first to last the Intendant gave me his sympathy and support. He could not alter laws, but he deprecated a "judaical" interpretation of them. His final letter to myself was as follows:—

"INTENDANCE ROYALE DE LA PROVINCE DE
FAUCIGNY, BONNEVILLE,
11 *Septembre* 1858.

"MONSIEUR,—J'apprends avec une véritable peine les difficultés que vous rencontrez de la part de M. le Guide Chef pour l'effectuation de votre périlleuse entreprise scientifique, mais je dois vous dire aussi avec regret que ces difficultés resident dans un réglement fait en vue de la sécurité des voyageurs, quelque puisse être le but de leurs excursions.

"Désireux néanmoins de vous être utile notamment en la circonstance, j'invite aujourd'hui même M. le Guide Chef à avoir égard à votre projet, à faire en sa faveur une exception au réglement ci devant eu, tant qu'il n'y aura aucun danger pour votre sûreté et celles des personnes qui vous accompagneront, et enfin de se prêter dans les limites de ses moyens et attributions pour l'heureux succès de l'expédition, dont les conséquences et résultats n'intéressent pas seulement la science, mais encore la vallée de Chamounix en particulier.

"Agréez, Monsieur, l'assurance de ma considération très-distinguée.

"Pour l'Intendant en congé, le Secrétaire,
"DELEGLISE."

While waiting for this permission I employed myself in various ways. On the 2nd of September I ascended

Second Ascent of Monte Rosa 153

the Brévent, from which Mont Blanc is seen to great advantage. From Chamouni its vast slopes are so foreshortened that one gets a very imperfect idea of the extent to be traversed to reach the summit. What, however, struck me most on the Brévent was the changed relation of the Aiguille du Dru and the Aiguille Verte. From Montanvert the former appears a most imposing mass, while the peak of the latter appears rather dwarfed behind it; but from the Brévent the Aiguille du Dru is a mere pinnacle stuck in the breast of the grander pyramid of the Aiguille Verte.

On the 4th I rose early, and, strapping on my telescope, ascended to the Montanvert, where I engaged a youth to accompany me up the glacier. The heavens were clear and beautiful—blue over the Aiguille du Dru, blue over the Jorasse and Mont Mallet, deep blue over the pinnacles of Charmoz, and the same splendid tint stretched grandly over the Col du Géant and its Aiguille. No trace of condensation appeared till towards eleven o'clock, when a little black balloon of cloud swung itself over the Aiguilles Rouges. At one o'clock there were two large masses and a little one between them; while higher up a white veil, almost too thin to be visible, spread over a part of the heavens. At the zenith, however, and south, north, and west, the blue seemed to deepen as the day advanced. I visited the ice-wall at the Tacul, which seemed lower than it was last year; the cascade of le Géant appeared also far less imposing. Only in the early part of summer do we see the ice in its true grandeur: its edges and surfaces are then sharp and clear, but afterwards its nobler masses shrink under the influence of sun and air. The *séracs* now appeared wasted and dirty, and not the sharp angular ice-castles which rose so grandly when I first saw them. Thirteen men had crossed the Col du Géant on the day previous, and left an ample trace behind them. This I followed nearly to the summit of the fall. The condition of the glacier was totally different from that of the opposite side on the previous year. The ice was riven, burrowed, and honeycombed, but the track amid all was easy: a vigorous English maiden might have ascended the fall without much difficulty. My object now was to examine the structure of the fall; but the ice was not in a good condition for such an examination: it was too much broken. Still a definite structure was in many places to be traced, and some of them apparently showed structure and bedding at a high angle to each other, but I could not be certain of it. I paused at every commanding point of view and examined the ice through my opera-glass; but the result was inconclusive. I observed that the terraces which compose the fall do not front the middle of the glacier, but turn their foreheads rather towards its eastern side, and the consequence is that the protuberances lower down, which are the remains of these terraces, are highest at the same side. Standing at the base of the Aiguille Noire, and looking downwards where the Glacier des Périades pushes itself against the Géant, a series of fine crumples is formed on the former, cut across by crevasses, on the walls of which a forward and backward dipping of the blue veins is exhibited. Huge crumples are also formed by the Glacier du Géant, which are well seen from a point nearly opposite the lowest lateral moraine of the Glacier des Périades. In some cases the upper portions of the crumples had scaled off so as to form arches of ice—a consequence doubtless of the pressure.

The beauty of some Alpine skies is treacherous; in fact the deepest blue often indicates an atmosphere charged almost to saturation with aqueous vapour. This was the case on the present occasion. Soon after reaching Chamouni in the evening, rain commenced and continued with scarcely any intermission until the afternoon of the 8th. I had given up all hopes of being able to ascend Mont Blanc; and hence resolved to place the thermometers in some more accessible position. On the 9th accordingly, accompanied by Mr. Wills, Balmat, and some other friends, I ascended to the summit of the Jardin, where we placed two thermometers: one in the ice, at a depth of three feet below the surface; another

Second Ascent of Monte Rosa

on a ledge of the highest rock.[1] The boiling-point of water at this place was 194.6° Fahr.

Deep snow was upon the Talèfre, and the surrounding precipices were also heavily laden. Avalanches thundered incessantly from the Aiguille Verte and the other mountains. Scarcely five minutes on an average intervened between every two successive peals; and after the direct shock of each avalanche had died away the air of the basin continued to be shaken by the echoes reflected from its bounding walls.

The day was far spent before we had completed our work. All through the weather had been fine, and towards evening augmented to magnificence. As we descended the glacier from the Couvercle the sun was just disappearing, and the western heaven glowed with crimson, which crept gradually up the sky until finally it reached the zenith itself. Such intensity of colouring is exceedingly rare in the Alps; and this fact, together with the known variations in the intensity of the firmamental blue, justify the conclusion that the colouring must, in a great measure, be due to some *variable constituent* of the atmosphere. If *the air* were competent to produce these magnificent effects they would be the rule instead of the exception.

No sooner had the thermometers been thus disposed of than the weather appeared to undergo a permanent change. On the 10th it was perfectly fine—not the slightest mist upon Mont Blanc; on the 11th this was also the case. Balmat still had the old thermometer to which I have already referred; it might not do to show the minimum temperature of the air, but it might show the temperature at a certain depth below the surface. I find in my own case that the finishing of work has a great moral value: work completed is a safe fulcrum for the performance of other work; and even though in the course of our labours experience should show us a better means of accomplishing a given end, it is often far preferable to reach the end,

---

[1] The minimum temperature of the subsequent winter, as shown by this thermometer, was −6° Fahr., or 38° below the freezing point. The instrument placed in the ice was broken.

even by defective means, than to swerve from our course. The habits which this conviction had superinduced no doubt influenced me when I decided on placing Balmat's thermometer on the summit of Mont Blanc.

SECOND ASCENT OF MONT BLANC, 1858

XXV

On the 12th of September, at 5½ A.M. the sunbeams had already fallen upon the mountain; but though the sky above him, and over the entire range of the Aiguilles, was without a cloud, the atmosphere presented an appearance of turbidity resembling that produced by the dust and thin smoke mechanically suspended in a London atmosphere on a dry summer's day. At 20 minutes past 7 we quitted Chamouni, bearing with us the good wishes of a portion of its inhabitants.

A lady accompanied us on horseback to the point where the path to the Grands Mulets deviates from that to the Plan des Aiguilles; here she turned to the left, and we proceeded slowly upwards, through woods of pine, hung with fantastic lichens: escaping from the gloom of these, we emerged upon slopes of bosky underwood, green hazel, and green larch, with the red berries of the mountain-ash shining brightly between them. Through the air above us, like gnomons of a vast sundial, the Aiguilles cast their fan-like shadows, which moved round as the day advanced. Slopes of rhododendrons with withered flowers next succeeded, but the colouring of the bilberry leaves was scarcely less exquisite than the freshest bloom of the Alpine rose. For a long time we were in the cool shadow of the mountain, catching, at intervals, through the twigs in front of us, glimpses of the sun surrounded by coloured spectra. On one occasion a brow rose in front of me; behind it was a lustrous space of heaven, adjacent to the sun, which, however, was hidden behind the brow; against this space the twigs and weeds upon the summit of the brow shone as if they were self-luminous, while some bits of thistle-down floating in the air appeared, where they crossed this portion of the heavens, like fragments of the sun himself. Once the orb appeared behind a rounded mass of snow which lay near the summit of the Aiguille du Midi. Looked at with the naked eyes, it seemed to possess a billowy motion, the light darting from it in dazzling curves,—a subjective effect produced by the abnormal action of the intense light upon the eye. As the sun's disk came more into view, its rays, however, still grazing the summit of the mountain, interference-spectra darted from it on all sides, and surrounded it with a glory of richly coloured bars. Mingling, however, with the grandeur of nature, we had the anger and obstinacy of man. With a view to subsequent legal proceedings, the Guide Chef sent a spy after us, who, having satisfied himself of our delinquency, took his unpleasant presence from the splendid scene.

Strange to say, though the luminous appearance of bodies projected against the sky adjacent to the rising sun is a most striking and beautiful phenomenon, it is hardly ever seen by either guides or travellers; probably because they avoid looking towards a sky the brightness of which is painful to the eyes. In 1859 Auguste Balmat had never seen the effect; and the only written description of it which we possess is one furnished by Professor Necker, in a letter to Sir David Brewster, which is so interesting that I do not hesitate to reproduce it here:—

"I now come to the point," writes M. Necker, "which you particularly wished me to describe to you; I mean the luminous appearance of trees, shrubs, and birds, when seen from the foot of a mountain a little before sunrise. The wish I had to see again the phenomenon before attempting to describe it made me detain this letter a few days, till I had a fine day to go to see it at the Mont Salève; so yesterday I went there, and studied the fact, and in elucidation of it I made a little drawing, of which I give you here a copy: it will, with the explanation and the annexed diagram (Fig. 9), impart to you, I hope, a correct idea of the phenomenon. You must conceive the observer placed at the foot of a hill interposed between him and the place where the sun is rising, and thus entirely in the shade; the upper margin of the mountain is covered with woods or detached trees and shrubs, which are projected as dark objects on a very bright and clear sky, except at the very place where the sun is just going to rise, for there all the trees and shrubs bordering the margin are entirely,—branches, leaves, stem and all,—of a pure and brilliant white, appearing extremely bright and luminous, although projected on a most brilliant and luminous sky, as that part of it which surrounds the sun always is. All the minutest details, leaves, twigs, &c., are most delicately preserved, and you would fancy you saw these trees and forests made of the purest silver, with all the skill of the most expert workman. The swallows and other birds flying in those particular spots appear like sparks of the most brilliant white. Unfortunately, all these details, which add so much to the beauty of this splendid phenomenon, cannot be represented in such small sketches.

"Neither the hour of the day nor the angle which the object makes with the observer appears to have any effect; for on some occasions I have seen the phenomenon take place at a very early hour in the morning. Yesterday it was 10 A.M., when I saw it as represented in Fig. 10. I saw it again on the same day at 5 P.M., at a different place of the same mountain, for which the sun was just setting. At one time the angle of elevation of the lighted white shrubs above the horizon of the spectator was about 20°, while at another place it was only 15°. But the extent of the field of illumination is variable, according to the distance at which the spectator is placed from it. When the object behind which the sun is just going to rise, or has just been setting, is very near, no such effect takes place. In the case represented in Fig. 9 the distance was about 194 mètres, or 636 English feet, from the spectator in a direct line, the height above his level being 60 mètres, or 197 English feet, and the horizontal line drawn from him to the horizontal projection of these points on the plane of his horizon being 160 mètres, or 525

Second Ascent of Mont Blanc 159

English feet, as will be seen in the following diagram, Fig. 10.

"In this case only small shrubs and the lower half of the stem of a tree are illuminated white, and the horizontal extent of this effect is also comparatively small; while at other places when I was near the edge behind which the sun was going to rise no such effect took place. But on the contrary, when I have witnessed the phenomenon at a greater distance and at a greater height, as I have seen it other times on the same and on other mountains of the Alps, large tracts of forests and immense spruce-firs were illuminated white throughout their whole length, as I have attempted to represent in

Fig. 9

Fig. 11, and the corresponding diagram, Fig. 12. Nothing can be finer than these silver-looking spruce-forests. At the same time, though at a distance of more than 1000 mètres, a vast number of large swallows or swifts (*Cypselus alpinus*), which inhabit these high rocks, were seen as small brilliant stars or sparks moving rapidly in the air.

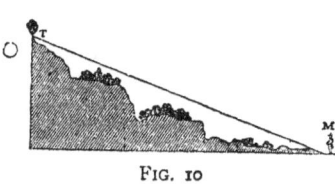

Fig. 10

From these facts it appears to me obvious that the extent of the illuminated spots varies in a direct ratio of their distance; but at the same time that there must be a constant angular space, corresponding probably to the zone, a few minutes of a degree wide, around the sun's disk, which is a limit to the occurrence of the appearance. This would explain how the real extent which it occupies on the earth's surface varies with the relative distance of the spot from the eye of the observer, and accounts also for the phenomenon being never seen in the low country, where I have often looked for it in vain. Now that you are acquainted with the circumstances of the fact, I have no doubt you will easily observe it in some part or other of your Scotch hills; it may be some long heather or furze will play the part of our Alpine forests, and I would advise you to try and place a beehive in the required position, and it would perfectly represent our swallows, sparks, and stars."

FIG. 11

Our porters, with one exception, reached the Pierre l'Echelle as soon as ourselves; and here having refreshed

FIG. 12

themselves, and the due exchange of loads having been made, we advanced upon the glacier, which we crossed until we came nearly opposite to the base of the Grands Mulets. The existence of one wide crevasse, which was deemed impassable, had this year introduced the practice

Second Ascent of Mont Blanc

of assailing the rocks at their base, and climbing them to the cabin, an operation which Balmat wished to avoid. At Chamouni, therefore, he had made inquiries regarding the width of the chasm, and acting on his advice I had had a ladder constructed in two pieces, which, united together by iron attachments, was supposed to be of sufficient length to span the fissure. On reaching the latter, the pieces were united, and the ladder thrown across, but the bridge was so frail and shaky at the place of junction, and the chasm so deep, that Balmat pronounced the passage impracticable.

The porters were all grouped beside the crevasse when this announcement was made, and, like hounds in search of the scent, the group instantly broke up, seeking in all directions for a means of passage. The talk was incessant and animating; attention was now called in one direction, anon in another, the men meanwhile throwing themselves into the most picturesque groups and attitudes. All eyes at length were directed upon a fissure which was spanned at one point by an arch of snow, certainly under two feet deep at the crown. A stout rope was tied round the waist of one of our porters, and he was sent forward to test the bridge. He approached it cautiously, treading down the snow to give it compactness, and thus make his footing sure as he advanced; bringing regelation into play, he gave the mass the necessary continuity, and crossed in safety. The rope was subsequently stretched over the *pont*, and each of us causing his right hand to slide along it, followed without accident. Soon afterwards, however, we met with a second and very formidable crevasse, to cross which we had but half of our ladder, which was applied as follows:—The side of the fissure on which we stood was lower than the opposite one; over the edge of the latter projected a cornice of snow, and a ledge of the same material jutted from the wall of the crevasse a little below us. The ladder was placed from ledge to cornice, both of its ends being supported by snow. I could hardly believe that so frail a bearing could possibly support a man's weight; but a porter was tied as before, and sent up the ladder, while we followed protected by the rope. We were afterwards tied together, and thus advanced in an orderly line to the Grands Mulets.

The cabin was wet and disagreeable, but the sunbeams fell upon the brown rocks outside, and thither Mr. Wills and myself repaired to watch the changes of the atmosphere. I took possession of the flat summit of a prism of rock, where, lying upon my back, I watched the clouds forming, and melting, and massing themselves together, and tearing themselves like wool asunder in the air above. It was nature's language addressed to the intellect; these clouds were visible symbols which enabled us to understand what was going on in the invisible air. Here unseen currents met, possessing different temperatures, mixing their contents both of humidity and motion, producing a mean temperature unable to hold their moisture in a state of vapour. The water-particles, obeying their mutual attractions, closed up, and a visible cloud suddenly shook itself out, where a moment before we had the pure blue of heaven. Some of the clouds were wafted by the air towards atmospheric regions already saturated with moisture, and along their frontal borders new cloudlets ever piled themselves, while the hinder portions, invaded by a drier or a warmer air, were dissipated; thus the cloud advanced, with gain in front and loss behind, its permanence depending on the balance between them. The day waned, and the sunbeams began to assume the colouring due to their passage through the horizontal air. The glorious light, ever deepening in colour, was poured bounteously over crags, and snows, and clouds, and suffused with gold and crimson the atmosphere itself. I had never seen anything grander than the sunset on that day. Clouds with their central portions densely black, denying all passage to the beams which smote them, floated westward, while the fiery fringes which bordered them were rendered doubly vivid by contrast with the adjacent gloom. The smaller and more attenuated clouds were intensely illuminated throughout. Across other inky masses were drawn zigzag bars of radiance which resembled streaks of lightning. The firmament between the clouds faded from a blood-red through orange and daffodil into an exquisite green, which spread like a sea of glory through which those magnificent argosies slowly sailed. Some of the clouds were drawn in straight chords across the arch of heaven, these being doubtless the sections of layers of cloud whose horizontal dimensions were hidden from us. The cumuli around and near the sun himself could not be gazed upon, until, as the day declined, they gradually lost their effulgence and became tolerable to the eyes. All was calm—but there was a wildness in the sky like that of anger, which boded evil passions on the part of the atmosphere. The sun at length sank behind the hills, but for some time afterwards carmine clouds swung themselves on high, and cast their ruddy hues upon the mountain snows. Duskier and colder waxed the west, colder and sharper the breeze of evening upon the Grands Mulets, and as twilight deepened towards night, and the stars commenced to twinkle through the chilled air, we retired from the scene.

The anticipated storm at length gave notice of its coming. The sea-waves, as observed by Aristotle, sometimes reach the shore before the wind which produces them is felt; and here the tempest sent out its precursors, which broke in detached shocks upon the cabin before the real storm arrived. Billows of air, in ever quicker succession, rolled over us with a long surging sound, rising and falling as crest succeeded trough and trough succeeded crest. And as the pulses of a vibrating body, when their succession is quick enough, blend to a continuous note, so these fitful gusts linked themselves finally to a storm which made its own wild music among the crags. Grandly it swelled, carrying the imagination out of doors, to the clouds and darkness, to the loosened avalanches and whirling snow upon the mountain heads. Moored to the rock on two sides, the cabin stood firm, and its manifest security allowed the mind the undisturbed enjoyment of the atmospheric war. We were powerfully shaken, but had no fear of being uprooted; and a certain grandeur of the heart rose responsive to the grandeur of the storm. Mounting higher and higher, it at length reached its maximum strength, from which it lowered fitfully, until at length, with a melancholy wail, it bade our rock farewell.

A little before half-past one we issued from the cabin. The night being without a moon, we carried three lanterns. The heavens were crowded with stars, among which, however, angry masses of cloud here and there still wandered. The storm, too, had left a rearguard behind it; and strong gusts rolled down upon us at intervals, at one time, indeed, so violent as to cause Balmat to express doubts of our being able to reach the summit. With a thick handkerchief bound around my hat and ears I enjoyed the onset of the wind. Once, turning my head to the left, I saw what appeared to me to be a huge mass of stratus cloud, at a great distance, with the stars shining over it. In another instant a precipice of *névé* loomed upon us; we were close to its base, and along its front the annual layers were separated from each other by broad dark bands. Through the gloom it appeared like a cloud, the lines of bedding giving to it the stratus character.

Immediately before lying down on the previous evening I had opened the little window of the cabin to admit some air. In the sky in front of me shone a curious nodule of misty light with a pale train attached to it. In 1853, on the side of the Brocken, I had observed, without previous notice, a comet discovered a few days previously by a former fellow student, and here was another "discovery" of the same kind. I inspected the stranger with my telescope, and assured myself that it was a comet. Mr. Wills chanced to be outside at the time, and made the same observation independently. As we now advanced up the mountain its ominous light gleamed behind us, while high up in heaven to our left the planet Jupiter burned like a lamp of intense brightness. The Petit Plateau forms a kind of reservoir for the avalanches of the Dôme du Goûté, and this year the accumulation of frozen débris upon it was enormous. We could see nothing but the ice-blocks on which the light of the lanterns immediately fell; we only knew that they had been discharged from the *séracs*, and that similar masses now rose threatening to our right, and might at any moment leap down upon us. Balmat commanded silence, and urged us to move across the plateau with all possible celerity. The warning of our guide, the wild and rakish appearance of the sky, the spent projectiles at our feet, and the comet with its "horrid hair" behind, formed a combination eminently calculated to excite the imagination.

And now the sky began to brighten towards dawn, with that deep and calm beauty which suggests the thought of adoration to the human mind. Helped by the contemplation of the brightening east, which seemed to lend lightness to our muscles, we cheerily breasted the steep slope up to the Grand Plateau. The snow here was deep, and each of our porters took the lead in turn. We paused upon the Grand Plateau and had breakfast; digging, while we halted, our feet deeply into the snow. Thence up to the corridor, by a totally different route from that pursued by Mr. Hirst and myself the year previously; the slope was steep, but it had not a precipice for its boundary. Deep steps were necessary for a time, but when we reached the summit our ascent became more gentle. The eastern sky continued to brighten, and by its illumination the Grand Plateau and its bounding heights were lovely beyond conception. The snow was of the purest white, and the glacier, as it pushed itself on all sides into the basin, was riven by fissures filled with a cœrulean light, which deepened to inky gloom as the vision descended into them. The edges were overhung with fretted cornices, from which depended long clear icicles, tapering from their abutments like spears of crystal. The distant fissures, across which the vision ranged obliquely without descending into them, emitted that magical firmamental shimmer which, contrasted with the pure white of the snow, was inexpressibly lovely. Near to us also grand castles of ice reared themselves, some erect, some overturned, with clear cut sides, striped by the courses of the annual snows, while high above the *séracs* of the plateau rose their still grander brothers of the Dôme du Gouté. There was a nobility in this glacier scene which I think I have never seen surpassed;—a strength of nature, and yet a tenderness, which at once raised and purified the soul. The gush of the direct sunlight could add nothing to this heavenly beauty; indeed I thought its yellow beams a profanation as they crept down from the humps of the Dromedary, and invaded more and more the solemn purity of the realm below.

Our way lay for a time amid fine fissures with blue walls, until at length we reached the edge of one which elicited other sentiments than those of admiration. It must be crossed. At the opposite side was a high and steep bank of ice which prolonged itself downwards, and ended in a dependent eave of snow which quite overhung the chasm, and reached to within about a yard of our edge of the crevasse. Balmat came forward with his axe, and tried to get a footing on the eave: he beat it gently, but the axe went through the snow, forming an aperture through which the darkness of the chasm was rendered visible. Our guide was quite free, without rope or any other means of security; he beat down the snow so as to form a kind of stirrup, and upon this he stepped. The stirrup gave way, it was right over the centre of the chasm, but with wonderful tact and coolness he contrived to get sufficient purchase from the yielding mass to toss himself back to the side of the chasm. The rope was now brought forward and tied round the waist of one of the porters; another step was cautiously made in the eave of snow, the man was helped across, and lessened his own weight by means of his hatchet. He gradually got footing on the face of the steep, which he mounted by escaliers; and on reaching a sufficient height he cut two large steps in which his feet might rest securely. Here he laid his breast against the sloping wall, and another person was sent forward, who drew himself up by the rope which was attached to the leader. Thus we all passed, each of us in turn bearing the strain of his successor upon the rope; it was our last difficulty, and

Second Ascent of Mont Blanc

we afterwards slowly plodded through the snow of the corridor towards the base of the Mur de la Côte.

Climbing zigzag, we soon reached the summit of the Mur, and immediately afterwards found ourselves in the midst of cold drifting clouds, which obscured everything. They dissolved for a moment and revealed to us the sunny valley of Chamouni; but they soon swept down again and completely enveloped us. Upon the Calotte, or last slope, I felt no trace of the exhaustion which I had experienced last year, but enjoyed free lungs and a quiet heart. The clouds now whirled wildly round us, and the fine snow, which was caught by the wind and spit bitterly against us, cut off all visible communication between us and the lower world. As we approached the summit the air thickened more and more, and the cold, resulting from the withdrawal of the sunbeams, became intense. We reached the top, however, in good condition, and found the new snow piled up into a sharp *arête*, and the summit of a form quite different from that of the *Dos d'un Ane*, which it had presented the previous year. Leaving Balmat to make a hole for the thermometer, I collected a number of batons, drove them into the snow, and, drawing my plaid round them, formed a kind of extempore tent to shelter my boiling-water apparatus. The covering was tightly held, but the snow was as fine and dry as dust, and penetrated everywhere: my lamp could not be secured from it, and half a box of matches was consumed in the effort to ignite it. At length it did flame up, and carried on a sputtering combustion. The cold of the snow-filled boiler condensing the vapour from the lamp gradually produced a drop, which, when heavy enough to detach itself from the vessel, fell upon the flame and put it out. It required much patience and the expenditure of many matches to relight it. Meanwhile the absence of muscular action caused the cold to affect our men severely. My beard and whiskers were a mass of clotted ice. The batons were coated with ice, and even the stem of my thermometer, the bulb of which was in hot water, was covered by a frozen enamel. The clouds whirled, and the little snow granules hit spitefully against the skin wherever it was exposed. The temperature of the air was 20° Fahr. below the freezing point. I was too intent upon my work to heed the cold much, but I was numbed; one of my fingers had lost sensation, and my right heel was in pain: still I had no thought of forsaking my observation until Mr. Wills came to me and said that we must return speedily, for Balmat's hands were *gelées*. I did not comprehend the full significance of the word; but, looking at the porters, they presented such an aspect of suffering that I feared to detain them longer. They looked like worn old men, their hair and clothing white with snow, and their faces blue, withered, and anxious-looking. The hole being ready, I asked Balmat for the magnet to arrange the index of the thermometer: his hands seemed powerless. I struck my tent, deposited the instrument, and, as I watched the covering of it up, some of the party, among whom were Mr. Wills and Balmat, commenced the descent.[1]

I followed them speedily. Midway down the Calotte I saw Balmat, who was about a hundred yards in advance of me, suddenly pause and thrust his hands into the snow, and commence rubbing them vigorously. The suddenness of the act surprised me, but I had no idea at the time of its real significance: I soon came up to him; he seemed frightened, and continued to beat and rub his hands, plunging them, at quick intervals, into the snow. Still I thought the thing would speedily pass away, for I had too much faith in the man's experience to suppose that he would permit himself to be seriously injured. But it did not pass as I hoped it would, and the terrible possibility of his losing his hands presented itself to me. He at length became exhausted by his own efforts, staggered like a drunken man, and fell upon the snow. Mr. Wills and myself took each a hand, and continued the process of beating and rubbing. I feared that we

---

[1] In August, 1859, I found the temperature of water, boiling in an open vessel at the summit of Mont Blanc, to be 184.95° Fahr. On that occasion also, though a laborious search was made for the thermometer, it could not be found.

should injure him by our blows, but he continued to exclaim, "N'ayez pas peur, frappez toujours, frappez fortement!" We did so, until Mr. Wills became exhausted, and a porter had to take his place. Meanwhile Balmat pinched and bit his fingers at intervals, to test their condition; but there was no sensation. He was evidently hopeless himself; and, seeing him thus, produced an effect upon me that I had not experienced since my boyhood—my heart swelled, and I could have wept like a child. The idea that I should be in some measure the cause of his losing his hands was horrible to me; schemes for his support rushed through my mind with the usual swiftness of such speculations, but no scheme could restore to him his lost hands. At length returning sensation in one hand announced itself by excruciating pain. "Je souffre!" he exclaimed at intervals —words which, from a man of his iron endurance, had a more than ordinary significance. But pain was better than death, and, under the circumstances, a sign of improvement. We resumed our descent, while he continued to rub his hands with snow and brandy, thrusting them at every few paces into the mass through which we marched. At Chamouni he had skilful medical advice, by adhering to which he escaped with the loss of six of his nails—his hands were saved.

I cannot close this recital without expressing my admiration of the dauntless bearing of our porters, and of the cheerful and efficient manner in which they did their duty throughout the whole expedition. Their names are Edouard Bellin, Joseph Favret, Michel Payot, Joseph Folliguet, and Alexandre Balmat.

XXVI

The hostility of the chief guide to the expedition was not diminished by the letter of the Intendant; and he at once entered a *procès verbal* against Balmat and his companions on their return to Chamouni. I felt that the power thus vested in an unlettered man to arrest the progress of scientific observations was so anomalous, that the enlightened and liberal Government of Sardinia would never tolerate such a state of things if properly represented to it. The British Association met at Leeds that year, and to it, as a guardian of science, my thoughts turned. I accordingly laid the case before the Association, and obtained its support: a resolution was unanimously passed " that application be made to the Sardinian authorities for increased facilities for making scientific observations in the Alps."

Considering the arduous work which Balmat had performed in former years in connection with the glaciers, and especially his zeal in determining, under the direction of Professor Forbes, their winter motion—for which, as in the case above recorded, he refused all personal remuneration—I thought such services worthy of some recognition on the part of the Royal Society. I suggested this to the Council, and was met by the same cordial spirit of co-operation which I had previously experienced at Leeds. A sum of five-and-twenty guineas was at once voted for the purchase of a suitable testimonial; and a committee, consisting of Sir Roderick Murchison, Professor Forbes, and myself, was appointed to carry the thing out. Balmat was consulted, and he chose a photographic apparatus, which, with a suitable inscription, was duly presented to him.

Thus fortified, I drew up an account of what had occurred at Chamouni during my last visit, accompanied by a brief statement of the changes which seemed desirable. This was placed in the hands of the President of the British Association, to whose prompt and powerful co-operation in this matter every Alpine explorer who aspires to higher ground than ordinary is deeply indebted. The following letter assured me that the facility applied for by the British Association would be granted by the Sardinian Government, and that future men of science would find in the Alps a less embarrassed field of operations than had fallen to my lot in the summer of 1858.

"12 HERTFORD STREET, MAYFAIR, W.,
February 18, 1859.

"MY DEAR SIR,—Having, as I informed you in my last note, communicated with the Sardinian Minister Plenipotentiary the day after receiving your statement relative to the guides at Chamouni, I have been favoured by replies from the Minister, of the 4th and 17th February. In the first the Marquis d'Azeglio assures me that he will bring the subject before the competent authorities at Turin, accompanying the transmission 'd'une récommendation toute spéciale.' In the second letter the Marquis informs me that 'the preparation of new regulations for the guides at Chamouni had for some time occupied the attention of the Minister of the Interior, and that these regulations will be in rigorous operation, in all probability, at the commencement of the approaching summer.' The Marquis adds that, 'as the regulations will be based upon a principle of much greater liberty, he has every reason to believe that they will satisfy all the desires of travellers in the interests of science.'

"With much pleasure at the opportunity of having been in any degree able to bring about the fulfilment of your wishes on the subject,—I remain, my dear Sir, faithfully yours,

"RICHARD OWEN,
"*Pres. Brit. Association.*

"Prof. TYNDALL, F.R.S."

It ought to be stated, that, previous to my arrival at Chamouni in 1858, an extremely cogent memorial drawn up by Mr. John Ball had been presented to the Marquis d'Azeglio by a deputation from the Alpine Club. It was probably this memorial which first directed the attention of the Sardinian Minister of the Interior to the subject.

WINTER EXPEDITION TO THE MER DE GLACE, 1859

XXVII

HAVING ten days at my disposal last Christmas, I was anxious to employ them in making myself acquainted with the winter aspects and phenomena of the Mer de Glace. On Wednesday, the 21st of December, I accordingly took my place to Paris, but on arriving at Folkestone found the sea so tempestuous that no boat would venture out.

The loss of a single day was more than I could afford, and this failure really involved the loss of two. Seeing, therefore, the prospect of any practical success so small, I returned to London, purposing to give the expedition up. On the following day, however, the weather lightened, and I started again, reaching Paris on Friday morning. On that day it was not possible to proceed beyond Macon, where, accordingly, I spent the night, and on the following day reached Geneva.

Much snow had fallen; at Paris it still cumbered the streets, and round about Macon it lay thick, as if a more than usually heavy cloud had discharged itself on that portion of the country. Between Macon and Roussillon it was lighter, but from the latter station onwards the quantity upon the ground gradually increased.

On Christmas morning, at 8 o'clock, I left Geneva by the diligence for Sallenches. The dawn was dull, but the sky cleared as the day advanced, and finally a dome of cloudless blue stretched overhead. The mountains were grand; their sunward portions of dazzling whiteness, while the shaded sides, in contrast with the blue sky behind them, presented a ruddy, subjective tint. The brightness of the day reached its maximum towards one o'clock, after which a milkiness slowly stole over the heavens, and increased in density until finally a drowsy turbidity filled the entire air. The distant peaks gradually blended with the white atmosphere above them and lost their definition. The black pine forests on the slopes of the mountains stood out in strong contrast to the snow; and, when looked at through the spaces enclosed by the tree branches at either side of the road, they appeared of a decided indigo-blue. It was only when thus detached by a vista in front that the blue colour was well seen, the air itself between the eye and the distant pines being the seat of the colour. Goethe would have regarded it as an excellent illustration of his 'Farbenlehre.'

We reached Sallenches a little after 4 P.M., where I endeavoured to obtain a sledge to continue my journey. A fit one was not to be found, and a carriage was therefore the only resort. We started at five; it was very dark, but the feeble reflex of the snow on each side of the road was preferred by the postilion to the light of lamps. Unlike the enviable ostrich, I cannot shut my eyes to danger when it is near: and as the carriage swayed towards the precipitous road side, I could not fold myself up, as it was intended I should, but, quitting the interior and divesting my limbs of every encumbrance, I took my seat beside the driver, and kept myself in readiness for the spring, which in some cases appeared imminent. My companion, however, was young, strong, and keen-eyed; and though we often had occasion for the exercise of the quality last mentioned, we reached Servoz without accident.

Here we baited, and our progress afterwards was slow and difficult. The snow on the road was deep and hummocky, and the strain upon the horses very great. Having crossed the Arve at the Pont-Pelissier, we both alighted, and I went on in advance. The air was warm, and not a whisper disturbed its perfect repose. There was no moon, and the heavy clouds, which now quite overspread the heavens, cut off even the feeble light of the stars. The sound of the Arve, as it rushed through the deep valley to my left, came up to me through crags and trees with a sad murmur. Sometimes on passing an obstacle, the sound was entirely cut off, and the consequent silence was solemn in the extreme. It was a churchyard stillness, and the tall black pines, which at intervals cast their superadded gloom upon the road, seemed like the hearse-plumes of a dead world. I reached a wooden hut, where a lame man offers batons, minerals, and *eau de vie*, to travellers in summer. It was forsaken, and half buried in the snow. I leaned against the door, and enjoyed for a time the sternness of the surrounding scene. My conveyance was far behind, and the intermittent tinkle of the horses' bells, which augmented instead of diminishing the sense of solitude, informed me of the progress and the pauses of the vehicle. At the summit of the road I halted until my companion reached me; we then both remounted, and proceeded slowly towards Les Ouches. We passed some houses, the aspect of which was even more dismal than that of Nature; their roofs were loaded with snow, and white buttresses were reared against the walls. There was no sound, no light, no voice of joy to indicate that it was the pleasant Christmas time. We once met the pioneer of a party of four drunken peasants: he came right against us, and the coachman had to pull up. Planting his feet in the snow and propping himself against the leader's shoulder, the bacchanal exhorted the postilion to drive on; the latter took him at his word, and overturned him in the snow. After this we encountered no living thing. The horses seemed seized by a kind of torpor, and leaned listlessly against each other; vainly the postilion endeavoured to rouse them by word and whip; they sometimes essayed to trot down the slopes, but immediately subsided to their former monotonous crawl. As we ascended the valley, the stillness of the air was broken at intervals by wild storm-gusts, sent down against us from Mont Blanc himself. These chilled me, so I quitted the carriage, and walked on. Not far from Chamouni, the road, for some distance, had been exposed to the full action of the wind, and the snow had practically erased it. Its left wall was completely covered, while a few detached stones, rising here and there above the surface, were the only indications of the presence and direction of the right-hand wall. I could not see the state of the surface, but I learned by other means that the snow had been heaped in oblique ridges across my path. I staggered over four or five of these in succession, sinking knee-deep, and finally found myself immersed to the waist. This made me pause; I thought I must have lost the road, and vainly endeavoured to check myself by the positions of surrounding objects. I turned back and met the carriage: it had stuck in one of the ridges; one horse was down, his hind legs buried to the haunches, his left fore leg plunged to the shoulder in snow, and the right one thrown forward upon the surface. *C'est bien la route?* demanded my companion. I went back exploring, and assured myself that we were over the road; but I recommended him to release the horses and leave the carriage to its fate. He, however, succeeded in extricating the leader, and while I went on in advance seeking out the firmer portions of the road, he followed, holding his horses by their heads; and half-an-hour's struggle of this kind brought us to Chamouni.

It also was a little "city of the dead." There was no living thing in the streets, and neither sound nor light in the houses. The fountain made a melancholy gurgle, one or two loosened window-shutters creaked harshly in the wind, and banged against the objects which limited their oscillations. The Hôtel de l'Union, so bright and gay in summer, was nailed up and forsaken; and the cross in front of it, stretching its snow-laden arms into the dim air, was the type of desolation. We rang the bell at the Hôtel Royal, but the bay of a watch-dog resounding through the house was long our only reply. The bell appeared powerless to wake the sleepers, and its sound mingled dismally with that of the wind howling through the deserted passages. The noise of my bootheel, exerted long on the front door, was at length effective; it was unbarred, and the physical heat of a good stove soon added itself to the warmth of the welcome with which my hostess greeted me.

December 26*th.*—The snow fell heavily, at frequent intervals, throughout the entire day. Dense clouds draped all the mountains, and there was not the least prospect of my being able to see across the Mer de Glace. I walked out alone in the dim light, and afterwards traversed the streets before going to bed. They were quite forsaken. Cold and sullen the Arve rolled under its wooden bridge, while the snow fell at intervals with heavy shock from the roofs of the houses, the partial echoes from the surfaces of the granules combining to render the sound loud and hollow. Thus the concerns of this little hamlet were changed and fashioned by the obliquity of the earth's axis, the chain of dependence which runs throughout creation, linking the roll of a planet alike with the interests of marmots and of men.

Tuesday, 27*th December.*—I rose at six o'clock, having arranged with my men to start at seven, if the weather at all permitted. Edouard Simond, my old assistant of 1857, and Joseph Tairraz were the guides of the party; the porters were Edouard Balmat, Joseph Simond (fils d'Auguste), François Ravanal, and another. They came at the time appointed; it was snowing heavily, and we agreed to wait till eight o'clock and then decide. They returned at eight, and finding them disposed to try the ascent to the Montanvert, it was not my place to baulk them. Through the valley the work was easy, as the snow had been partially beaten down, but we soon passed the habitable limits, and had to break ground for ourselves. Three of my men had tried to reach the Montanvert by *la Filia* on the previous Thursday, but their experience of the route had been such as to deter them from trying it again. We now chose the ordinary route, breasting the slope until we reached the cluster of chalets, under the projecting eave of one of which the men halted and applied "pattens" to their feet. These consisted of planks about sixteen inches long and ten wide, which were firmly strapped to the feet. My first impression was that they were worse than useless, for though they sank

Expedition to the Mer de Glace

less deeply than the unarmed feet, on being raised they carried with them a larger amount of snow, which, with the leverage of the leg, appeared to necessitate an enormous waste of force. I stated this emphatically, but the men adhered to their pattens, and before I reached the Montanvert I had reason to commend their practice as preferable to my theory. I was, however, guided by the latter, and wore no pattens. The general depth of the snow along the track was over three feet; the footmarks of the men were usually rigid enough to bear my weight, but in many cases I went through the crust which their pressure had produced, and sank suddenly in the mass. The snow became softer as we ascended, and my immersions more frequent, but the work was pure enjoyment, and the scene one of extreme beauty. The previous night's snow had descended through a perfectly still atmosphere, and had loaded all the branches of the pines; the long arms of the trees drooped under the weight, and presented at their extremities the appearance of enormous talons turned downwards. Some of the smaller and thicker trees were almost entirely covered, and assumed grotesque and beautiful forms; the upper part of one in particular resembled a huge white parrot with folded wings and drooping head, the slumber of the bird harmonising with the torpor of surrounding nature. I have given a sketch of it in Fig. 13.

FIG. 13

Previous to reaching the half-way spring, where the peasant girls offer strawberries to travellers in summer, we crossed two large couloirs filled with the débris of avalanches which had fallen the night before. Between

these was a ridge forty or fifty yards wide on which the
snow was very deep, the slope of the mountain also adding
a component to the fair thickness of the snow. My
shoulder grazed the top of the embankment to my right
as I crossed the ridge, and once or twice I found myself
waist deep in a vertical shaft from which it required a
considerable effort to escape. Suddenly we heard a deep
sound resembling the dull report of a distant gun, and at
the same moment the snow above us broke across, forming
a fissure parallel to our line of march. The layer of snow
had been in a state of strain, which our crossing brought
to a crisis: it gave way, but having thus relieved itself
it did not descend. Several times during the ascent the
same phenomenon occurred. Once, while engaged upon
a very steep slope, one of the men cried out to the
leader, "*Arrêtez!*" Immediately in front of the latter the
snow had given way, forming a zigzag fissure across the
slope. We all paused, expecting to see an avalanche
descend. Tairraz was in front; he struck the snow with
his baton to loosen it, but seeing it indisposed to descend
he advanced cautiously across it, and was followed by
the others. I brought up the rear. The steepness of
the mountain side at this place, and the absence of any
object to which one might cling, would have rendered
a descent with the snow in the last degree perilous, and
we all felt more at ease when a safe footing was secured
at the further side of the incline.

At the spring, which showed a little water, the men
paused to have a morsel of bread. The wind had changed,
the air was clearing, and our hopes brightening. As we
ascended the atmosphere went through some extraordinary
mutations. Clouds at first gathered round the Aiguille
and Dôme du Gouté, casting the lower slopes of the
mountain into intense gloom. After a little time all this
cleared away, and the beams of the sun striking detached
pieces of the slopes and summits produced an extraordinary effect. The Aiguille and Dôme were most singularly illuminated, and to the extreme left rose the white
conical hump of the Dromedary, from which a long
streamer of snow-dust was carried southward by the wind.

Expedition to the Mer de Glace

The Aiguille du Dru, which had been completely mantled during the earlier part of the day, now threw off its cloak of vapour and rose in most solemn majesty before us; half of its granite cone was warmly illuminated, and half in shadow. The wind was high in the upper regions, and, catching the dry snow which rested on the asperities and ledges of the Aiguille, shook it out like a vast banner in the air. The changes of the atmosphere, and the grandeur which they by turns revealed and concealed, deprived the ascent of all weariness. We were usually flanked right and left by pines, but once between the fountain and the Montanvert we had to cross a wide unsheltered portion of the mountain which was quite covered with the snow of recent avalanches. This was lumpy and far more coherent than the undisturbed snow. We took advantage of this, and climbed zigzag over the avalanches for three-quarters of an hour, thus reaching the opposite pines at a point considerably higher than the path. This, though not the least dangerous, was the least fatiguing part of the ascent.

I frequently examined the colour of the snow: though fresh, its blue tint was by no means so pronounced as I have seen it on other occasions; still it was beautiful. The colour is, no doubt, due to the optical reverberations which occur within a fissure or cavity formed in the snow. The light is sent from side to side, each time plunging a little way into the mass; and being ejected from it by reflection, it thus undergoes a sifting process, and finally reaches the eye as blue light. The presence of any object which cuts off this cross-fire of the light destroys the colour. I made conical apertures in the snow, in some cases three feet deep, a foot wide at the mouth, and tapering down to the width of my baton. When the latter was placed along the axis of such a cone, the blue light which had previously filled the cavity disappeared; on the withdrawal of the baton it was followed by the light, and thus by moving the staff up and down its motions were followed by the alternate appearance and extinction of the light. I have said that the holes made in the snow seemed filled with a blue light, and it certainly appeared as if the air contained in the cavities had itself been coloured, and thereby rendered visible, the vision plunging into it as into a blue medium. Another fact is perhaps worth notice: snow rarely lies so smooth as not to present little asperities at its surface; little ridges or hillocks, with little hollows between them. Such small hollows resemble, in some degree, the cavities which I made in the snow, and from them, in the present instance, a delicate light was sent to the eye, faintly tinted with the pure blue of the snow-crystals. In comparison with the spots thus illuminated, the little protuberances were grey. The portions most exposed to the light seemed least illuminated, and their defect in this respect made them appear as if a light-brown dust had been strewn over them.

After five hours and a half of hard work we reached the Montanvert. I had often seen it with pleasure. Often, having spent the day alone amid the *séracs* of the Col du Géant, on turning the promontory of Trélaporte on my way home, the sight of the little mansion has gladdened me, and given me vigour to scamper down the glacier, knowing that pleasant faces and wholesome fare were awaiting me. This day, also, the sight of it was most welcome, despite its desolation. The wind had swept round the auberge, and carried away its snow-buttresses, piling the mass thus displaced against the adjacent sheds, to the roofs of which one might step from the surface of the snow. The floor of the little château in which I lodged in 1857 was covered with snow, and on it were the fresh footmarks of a little animal—a marmot might have made such marks, had not the marmots been all asleep—what the creature was I do not know.

In the application of her own principles, Nature often transcends the human imagination; her acts are bolder than our predictions. It is thus with the motion of glaciers; it was thus at the Montanvert on the day now referred to. The floors, even where the windows appeared well closed, were covered with a thin layer of fine snow; and some of the mattresses in the bedrooms were coated to the depth of half an inch with this fine powder. Given a chink through which the finest dust can pass, dry snow appears competent to make its way through the same fissure. It had also been beaten against the windows, and clung there like a ribbed drapery. In one case an effect so singular was exhibited, that I doubted my eyes when I first saw it. In front of a large pane of glass, and quite detached from it, save at its upper edge, was a festooned curtain formed entirely of minute ice crystals. It appeared to be as fine as muslin; the ease of its curves and the depth of its folds being such as could not be excelled by the intentional arrangement of ordinary gauze. The frost-figures on some of the window-panes were also of the most extraordinary character: in some cases they extended over large spaces, and presented the appearance which we often observe in London; but on other panes they occurred in detached clusters, or in single flowers, these grouping themselves together to form miniature bouquets of inimitable beauty. I placed my warm hand against a pane which was covered by the crystallisation, and melted the frost-work which clung to it. I then withdrew my hand and looked at the film of liquid through a pocket-lens. The glass cooled by contact with the air, and after a time the film commenced to move at one of its edges; atom closed with atom, and the motion ran in living lines through the pellicle, until finally the entire film presented the beauty and delicacy of an organism. The connection between such objects and what we are accustomed to call the feelings may not be manifest, but it is nevertheless true that, besides appealing to the pure intellect of man, these exquisite productions can also gladden his heart and moisten his eyes.

The glacier excited the admiration of us all: not as in summer, shrunk and sullied like a spent reptile, steaming under the influence of the sun, its frozen muscles were compact; strength and beauty were associated in its aspect. At some places it was pure and smooth; at others frozen fins arose from it, high, steep, and sharply crested. Down the opposite mountain side arrested streams set themselves erect in successive terraces, the fronts of which were fluted pillars of ice. There was no sound of water; even the Nant Blanc, which gushes from a spring, and which some describe as permanent throughout the winter, showed no trace of existence. From the Montanvert to Trélaporte the Mer de Glace was all in shadow; but the sunbeams pouring down the corridor of the Géant ruled a beam of light across the glacier at its upper portion, smote the base of the Aiguille du Moine, and flooded the mountain with glory to its crest. At the opposite side of the valley was the Aiguille du Dru, with a banneret of snow streaming from its mighty cone. The Grande Jorasse, and the range of summits between it and the Aiguille du Géant, were all in view, and the Charmoz raised its precipitous cliffs to the right, and pierced with its splinter-like pinnacles the clear cold air. As the night drew on, the mountains seemed to close in upon us; and on looking out before retiring to rest, a scene so solemn had never before presented itself to my eyes or affected my imagination.

My men occupied the afternoon of the day of our arrival in making a preliminary essay upon the glacier while I prepared my instruments. To the person whom I intended to fix my stations, three others were attached by sound ropes of considerable length. Hidden crevasses we knew were to be encountered, and we had made due preparation for them. Throughout the afternoon the weather remained fine, and at night the stars shone out, but still with a feeble lustre. I could notice a turbidity gathering in the air over the range of the Brévent, which seemed disposed to extend itself towards us. At night I placed a chair in the middle of the snow, at some distance from the house, and laid on it a registering thermometer. A bountiful fire of pine logs was made in the *salle à manger;* a mattress was placed with its foot towards the fire, its middle line bisecting the right angle in which the fireplace stood; this being found by experiment to be the position in which the draughts from the door and from the windows most effectually neutralised each other. In this region of calms I lay down, and covering myself with blankets and duvets, listened to the crackling of the logs, and watched their ruddy flicker upon the walls, until I fell asleep.

The wind rose during the night, and shook the windows: one pane in particular seemed set in unison to the gusts, and responded to them by a loud and melodious vibration. I rose and wedged it round with *sous* and penny pieces, and thus quenched its untimely music.

December 28*th.* — We were up before the dawn. Tairraz put my fire in order, and I then rose. The temperature of the room at a distance of eight feet from the fire was two degrees of Centigrade below zero; the lowest temperature outside was eleven degrees of Centigrade below zero,—not at all an excessive cold. The clouds indeed had, during the night, thrown vast diaphragms across the sky, and thus prevented the escape of the earth's heat into space.

While my assistants were preparing breakfast I had time to inspect the glacier and its bounding heights. On looking up the Mer de Glace, the Grande Jorasse meets the view, rising in steep outline from the wall of cliffs which terminates the Glacier de Léchaud. Behind this steep ascending ridge, and upon it, a series of clouds had ranged themselves, stretching lightly along the ridge at some places, and at others collecting into ganglia. A string of rosettes was thus formed which were connected together by gauzy filaments. The portion of the heavens behind the ridge was near the domain of the rising sun, and when he cleared the horizon his red light fell upon the clouds, and ignited them to ruddy flames. Some of the lighter clouds doubled round the summit of the mountain, and swathed its black crags with a vestment of transparent red. The adjacent sky wore a strange and supernatural air; indeed there was something in the whole scene which baffled analysis, and the words of Tennyson rose to my lips as I gazed upon it:—

"God made Himself an awful rose of dawn."

I have spoken several times of the cloud-flag which the wind wafted from the summit of the Aiguille du Dru. On the present occasion this grand banner reached extraordinary dimensions. It was brindled in some places as if whipped into curds by the wind; but through these continuous streamers were drawn, which were bent into sinuosities resembling a waving flag at a masthead. All this was now illuminated with the sun's red rays, which also fringed with fire the exposed edges and pinnacles both of the Aiguille du Dru and the Aiguille Verte. Thus rising out of the shade of the valley the mountains burned like a pair of torches, the flames of which were blown half a mile through the air. Soon afterwards the summits of the Aiguilles Rouges were illuminated, and day declared itself openly among the mountains.

But these red clouds of the morning, magnificent though they were, suggested thoughts which tended to qualify the pleasure which they gave: they did not indicate good weather. Sometimes, indeed, they had to fight with denser masses, which often prevailed, swathing the mountains in deep neutral tint, but which, again yielding, left the glory of the sunrise augmented by contrast with their gloom. Between 8 and 9 A.M. we commenced the setting out of our first line, one of whose termini was a point about a hundred yards higher up than the Montanvert Hotel; a withered pine on the opposite mountain side marking the other terminus. The stakes made use of were four feet long. With the self-same baton which I had employed upon the Mer de Glace in 1857, and which Simond had preserved, the worthy fellow now took up the line. At some places the snow was very deep, but its lower portions were sufficiently compact to allow of a stake being firmly fixed in it. At those places where the wind had removed the snow or rendered it thin, the ice was pierced with an auger and the stake driven into it. The greatest caution was of course necessary on the part of the men; they were in the midst of concealed crevasses, and sounding was essential at every step. By degrees they withdrew from me, and approached the eastern boundary of the glacier, where the ice was greatly dislocated, and the labour of wading through the snow enormous. Long detours were sometimes necessary to reach a required point; but they were all accomplished, and we at length succeeded in fixing eleven stakes along this line, the most distant of which was within about eighty yards of the opposite side of the glacier.

The men returned, and I consulted them as to the possibility of getting a line across at the *Ponts;* but this was judged to be impossible in the time. We thought, however, that a second line might be staked out at some distance below the Montanvert. I took the theodolite down the mountain-slope, wading at times breast-deep in snow, and having selected a line, the men tied themselves together as before, and commenced the staking out. The work was slowly, but steadily and steadfastly done. The air darkened; angry clouds gathered around the mountains, and at times the glacier was swept by wild squalls. The men were sometimes hidden from me by the clouds of snow which enveloped them, but between those intermittent gusts there were intervals of repose, which enabled us to prosecute our work. This line was more difficult than the first one; the glacier was broken into sharp-edged chasms; the ridges to be climbed were steep, and the snow which filled the depressions profound. The oblique arrangement of the crevasses also magnified the labour by increasing the circuits. I saw the leader of the party often shoulder-deep in snow, treading the soft mass as a swimmer walks in water, and I felt a wish to be at his side to cheer him and to share his toil. Each man there, however, knew my willingness to do this if occasion required it, and wrought contented. At length the last stake being fixed, the faces of the men were turned homeward. The evening became wilder, and the storm rose at times to a hurricane. On the more level portions of the glacier the snow lay deep and unsheltered; among its frozen waves and upon its more dislocated portions it had been partially engulfed, and the residue was more or less in shelter. Over the former spaces dense clouds of snow rose, whirling in the air and cutting off all view of the glacier. The whole length of the Mer de Glace was thus divided into clear and cloudy segments, and presented an aspect of wild and wonderful turmoil. A large pine stood near me, with its lowest branch spread out upon the surface of the snow; on this branch I seated myself, and, sheltered by the trunk, waited until I saw my men in safety. The wind caught the branches of the trees, shook down their loads of snow, and tossed it wildly in the air. Every mountain gave a quota to the storm. The scene was one of most impressive grandeur, and the moan of the adjacent pines chimed in noble harmony with the picture which addressed the eyes.

At length we all found ourselves in safety within doors. The windows shook violently. The tempest was, however, intermittent throughout, as if at each effort it had exhausted itself, and required time to recover its strength. As I heard its heralding roar in the gullies of the mountains, and its subsequent onset against our habitation, I thought wistfully of my stations, not knowing whether they would be able to retain their positions in the face of such a blast. That night, however, as if the storm had sung our lullaby, we all slept profoundly, having arranged to commence our measurements as early as light permitted on the following day.

Thursday, 29th December.—" Snow, heavy snow: it must have descended throughout the entire night; the quantity freshly fallen is so great; the atmosphere at seven o'clock is thick with the descending flakes." At eight o'clock it cleared up a little, and I proceeded to my station, while the men advanced upon the glacier; but I had scarcely fixed my theodolite when the storm recommenced. I had a man to clear away the snow and otherwise assist me; he procured an old door from the hotel, and by rearing it upon its end sheltered the object-glass of the instrument. Added to the flakes descending from the clouds was the spitting snow-dust raised by the wind, which for a time so blinded me that I was unable to see the glacier. The measurement of the first stake was very tedious, but practice afterwards enabled me to take advantage of the brief lulls and periods of partial clearness with which the storm was interfused.

At nine o'clock my telescope happened to be directed upon the men as they struggled through the snow; all evidence of the deep track which they had formed yesterday having been swept away. I saw the leader sink and suddenly disappear. He had stood over a concealed fissure, the roof of which had given way and he had dropped in. I observed a rapid movement on the part of the remaining three men: they grouped themselves beside the fissure, and in a moment the missing man was drawn from between its jaws. His disappearance and appearance were both extraordinary. We had, as I have stated, provided for contingencies of this kind, and the man's rescue was almost immediate.

My attendant brought two poles from the hotel which we thrust obliquely into the snow, causing the free ends to cross each other; over these a blanket was thrown, behind which I sheltered myself from the storm as the men proceeded from stake to stake. At 9.30 the storm was so thick that I was unable to see the men at the stake which they had reached at the time; the flakes sped wildly in their oblique course across the field of the telescope. Some time afterwards the air became quite still, and the snow underwent a wonderful change. Frozen flowers similar to those I had observed on Monte Rosa fell in myriads. For a long time the flakes were wholly composed of these exquisite blossoms entangled together. On the surface of my woollen dress they were soft as down; the snow itself on which they fell seemed covered by a layer of down; while my coat was completely spangled with six-rayed stars. And thus prodigal Nature rained down beauty, and had done so here for ages unseen by man. And yet some flatter themselves with the idea that this world was planned with strict reference to human use; that the lilies of the field exist simply to appeal to the sense of the beautiful in man. True, this result is secured, but it is one of a thousand all equally important in the eyes of Nature. Whence those frozen blossoms? Why for æons wasted? The question reminds one of the poet's answer when asked whence was the Rhodora:—

"Why wert thou there, O rival of the rose?
   I never thought to ask, I never knew;
But in my simple ignorance suppose
   The self-same power that brought me there brought you!"[1]

---

[1] Emerson.

I sketched some of the crystals, but, instead of reproducing these sketches, which were rough and hasty, I have annexed two of the forms drawn with so much skill and patience by Mr. Glaisher.

We completed the measurement of the first line before eleven o'clock, and I felt great satisfaction in the thought that I possessed something of which the weather could not deprive me. As I closed my note-book and shifted the instrument to the second station, I felt that my expedition was already a success.

At a quarter past eleven I had my theodolite again fixed, and ranging the telescope along the line of pickets, I saw them all standing. Crossing the ice wilderness, and suggesting the operation of intelligence amid that scene of desolation, their appearance was pleasant to me. Just before I commenced, a solitary jay perched upon the summit of an adjacent pine and watched me. The air was still at the time, and the snow fell heavily. The flowers, moreover, were magnificent, varying from about the twentieth of an inch to two lines in diameter, while, falling through the quiet air, their forms were perfect. Adjacent to my theodolite was a stump of pine, from which I had the snow removed, in order to have something to kick my toes against when they became cold; and on the stump was placed a blanket to be used as a screen in case of need. While I remained at the station a layer of snow an inch thick fell upon this blanket, the whole layer being composed of these exquisite flowers. The atmosphere also was filled with them. From the clouds to the earth Nature was busy marshalling her atoms, and putting to shame by the beauty of her structures the comparative barbarities of Art.

My men at length reached the first station, and the measurement commenced. The storm drifted up the valley, thickening all the air as it approached. Denser and denser the flakes fell; but still, with care and tact I was able to follow my party to a distance of 800 yards. I had not thought it possible to see so far through so dense a storm. At this distance also my voice could be heard, and my instructions understood; for once, as the

Fig. 14

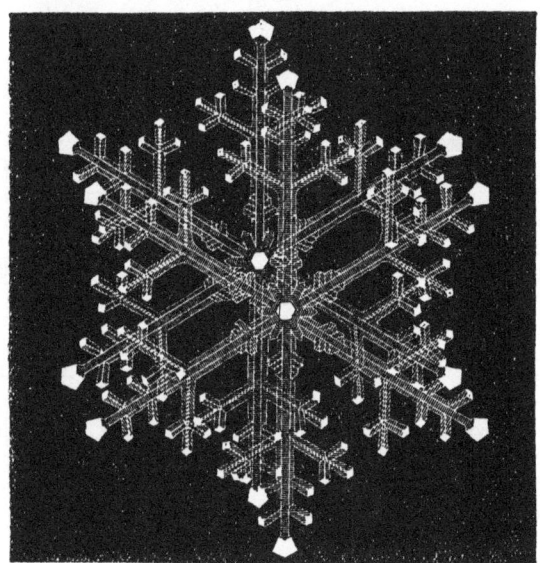

Fig. 15

man who took up the line stood behind his baton and prevented its projection against the white snow, I called out to him to stand aside, and he promptly did so. Throughout the entire measurement the snow never ceased falling, and some of the illusions which it produced were extremely singular. The distant boundary of the glacier appeared to rise to an extraordinary height, and the men wading through the snow appeared as if climbing up a wall. The labour along this line was still greater than on the former; on the steeper slopes especially the toil was great; for here the effort of the leader to lift his own body added itself to that of cutting his way through the snow. His footing I could see often yielded, and he slid back, checking his recession, however, by still plunging forward; thus, though the limbs were incessantly exerted, it was, for a time, a mere motion of vibration without any sensible translation. At the last stake the men shouted, "*Nous sommes finis!*" and I distinctly heard them through the falling snow. By this time I was quite covered with the crystals which clung to my wrapper. They also formed a heap upon my theodolite, rising over the spirit-levels and embracing the lower portion of the vertical arc. The work was done; I struck my theodolite and ascended to the hotel; the greatest depth of snow through which I waded reaching, when I stood erect, to within three inches of my breast.

The men returned; dinner was prepared and consumed; the disorder which we had created made good; the rooms were swept, the mattresses replaced, and the shutters fastened, where this was possible. We locked up the house, and with light hearts and lithe limbs commenced the descent. My aim now was to reach the source of the Arveiron, to examine the water and inspect the vault. With this view we went straight down the mountain. The inclinations were often extremely steep, and down these we swept with an avalanche-velocity; indeed usually accompanied by an avalanche of our own creation. On one occasion Balmat was for a moment overwhelmed by the descending mass: the guides were startled, but he emerged instantly. Tairraz followed him, and I followed Tairraz, all of us rolling in the snow at the bottom of the slope as if it were so much flour. My practice on the Finsteraarhorn rendered me at home here. One of the porters could by no means be induced to try this flying mode of descent. Simond carried my theodolite box, tied upon a crotchet on his back; and once, while shooting down a slope, he incautiously allowed a foot to get entangled; his momentum rolled him over and over down the incline, the theodolite emerging periodically from the snow during his successive revolutions. A succession of *glissades* brought us with amazing celerity to the bottom of the mountain, whence we picked our way amid the covered boulders and over the concealed arms of the stream to the source of the Arveiron.

The quantity of water issuing from the vault was considerable, and its character that of true glacier water. It was turbid with suspended matter, though not so turbid as in summer; but the difference in force and quantity would, I think, be sufficient to account for the greater summer turbidity. This character of the water could only be due to the grinding motion of the glacier upon its bed; a motion which seems not to be suspended even in the depth of winter. The temperature of the water was the tenth of a degree Centigrade above zero; that of the ice was half a degree below zero: this was also the temperature of the air, while that of the snow, which in some places covered the ice-blocks, was a degree and a quarter below zero.

The entrance to the vault was formed by an arch of ice which had detached itself from the general mass of the glacier behind: between them was a space through which we could look to the sky above. Beyond this the cave narrowed, and we found ourselves steeped in the blue light of the ice. The roof of the inner arch was perforated at one place by a shaft about a yard wide, which ran vertically to the surface of the glacier. Water had run down the sides of this shaft, and, being re-frozen below, formed a composite pillar of icicles at least twenty feet high and a yard thick, stretching quite from roof to floor. They were all united to a common surface at one side, but at the other they formed a series of flutings of exceeding beauty. This group of columns was bent at its base as if it had yielded to the forward motion of the glacier, or to the weight of the arch overhead. Passing over a number of large ice-blocks which partially filled the interior of the vault, we reached its extremity, and here found a sloping passage with a perfect arch of crystal overhead, and leading by a steep gradient to the air above. This singular gallery was about seventy feet long, and was floored with snow. We crept up it, and from the summit descended by a glissade to the frontal portion of the cavern. To me this crystal cave, with the blue light glistening from its walls, presented an aspect of magical beauty. My delight, however, was tame compared with that of my companions.

Looking from the blue arch westwards, the heavens were seen filled by crimson clouds, with fiery outliers reaching up to the zenith. On quitting the vault I turned to have a last look at those noble sentinels of the Mer de Glace, the Aiguille du Dru, and the Aiguille Verte. The glacier below the mountains was in shadow, and its frozen precipices of a deep cold blue. From this, as from a basis, the mountain cones sprang steeply heavenward, meeting half-way down the fiery light of the sinking sun. The right-hand slopes and edges of both pyramids burned in this light, while detached protuberant masses also caught the blaze, and mottled the mountains with effulgent spaces. A range of minor peaks ran slanting downwards from the summit of the Aiguille Verte; some of these were covered with snow, and shone as if illuminated with the deep crimson of a strontian flame. I was absolutely struck dumb by the extraordinary majesty of this scene, and watched it silently till the red light faded from the highest summits. Thus ended my winter expedition to the Mer de Glace.

Next morning, starting at three o'clock, I was driven by my two guides in an open sledge to Sallenches. The rain was pitiless and the road abominable. The distance, I believe, is only six leagues, but it took us five hours to

Expedition to the Mer de Glace

accomplish it. The leading mule was beyond the reach of Simond's whip, and proved a mere obstructive; during part of the way it was unloosed, tied to the sledge, and dragged after it. Simond afterwards mounted the hindmost beast and brought his whip to bear upon the leader; the jerking he endured for an hour and a half seemed almost sufficient to dislocate his bones. We reached Sallenches half-an-hour late, but the diligence was behind its time by this exact interval. We met it on the Pont St. Martin, and I transferred myself from the sledge to the interior. This was the morning of the 30th of December, and on the evening of the 1st of January I was in London.

I cannot finish this recital without saying one word about my men. Their behaviour was admirable throughout. The labour was enormous, but it was manfully and cheerfully done. I know Simond well; he is intelligent, truthful, and affectionate, and there is no guide of my acquaintance for whom I have a stronger regard. Joseph Tairraz is an extremely intelligent and able guide, and on this trying occasion proved himself worthy of my highest praise and commendation. Their two companions upon the glacier, Edouard Balmat (le Petit Balmat) and Joseph Simond (fils d'Auguste) acquitted themselves admirably; and it also gives me pleasure to bear testimony to the willing and efficient service of François Ravanal, who attended upon me during the observations.

MOUNTAINEERING IN 1861

The mottoes taken from Mr. Tennyson will be recognised by everybody:—the others are from the poems of Mr. Emerson.

MOUNTAINEERING IN 1861

A VACATION TOUR

ADDRESSED TO X

CHAPTER I

LONDON TO MEYRINGEN

> ' The mountain cheer, the frosty skies,
> Breed purer wits, inventive eyes;
> And then the moral of the place
> Hints summits of heroic grace.
> Men in these crags a fastness find
> To fight corruption of the mind,
> The insanity of towns to stem,
> With simpleness for stratagem.'

HERE I am at length, my friend, far away from the smoke and roar of London, with a blue sky bending over me, and the Rhine spreading itself in glimmering sheets beneath my window. Swift and silent the flashing river runs; not a whisper it utters here, but higher up it gets entangled in the props of a bridge and breaks into foam; its compressed bubbles snap like elastic springs, and shake the air into sonorous vibrations. Thus the rude mechanical motion of the river is converted into music. From the windows of the edifices along the banks gleam a series of reflected suns, each surrounded by a coloured glory. The hammer of the boat-builder rings on his plank, the leaves of the poplars rustle in the breeze, the watch-dog's honest bark is heard in the distance, and the current of Swiss life is poured like that of electricity in two directions across the bridge.

The scene is very tranquil; and the peace of the present is augmented by its contrast with the tumult of the past. Yesterday I travelled from Paris, and the day previous from London, when the trail of a spent storm swept across the sea and kept its anger awake. The stern of our boat went up and down, the distant craft were equally pendulous, and the usual results followed. Men's faces waxed green; roses faded from ladies' cheeks; while poor unconscious children yelled intermittently in the grasp of the demon which had taken possession of them. One rare pale maiden sat right in the line of the spray which was churned up by the paddle-wheel, and carried by the wind across the deck: she drew her shawl around her, and bore the violence of the ocean with the resignation of an angel; a white arm could be seen shining through the translucent muslin, but even against it the cruel brine beat as if it were a mere seaweed. I sat at rest, hovering fearfully on the verge of that doleful region, whose bourne most of those on board had already passed. A friend whom I accompanied betook himself early to the cabin, and there endured the tortures of the condemned. Nothing, perhaps, takes down the boasted supremacy of the human will more effectually than the smell and shiver of a steamer superposed upon the motion of the sea. We finally reached Boulogne, and sought to reconstitute our shattered energies at the restauration. The success was but partial. The soup was poor, and the *filets* reminded one of the reindeer boots of the Laplander, which their owner gnaws when other provisions fail. To one who regards physical existence as the mystic substratum of man's moral nature, few seem more ripe for judgment than he who debases that nature by the ministration of unwholesome food. The self-same atmosphere forced through one instrument produces music; through another, noise: and thus the spirit of life, acting through the human organism, is rendered demoniac or angelic by the health or the disease which originate in what we eat.

The morning of the 1st of August finds us on our way from Paris to Bale. The heavens are unstained by cloud, and as the day advances the sunbeams grow stronger, and are drunk in with avidity by the absorbent cushions which surround us. In addition to this source of temperature, eight human beings, each burning the slow fire which we call life, are cooped within the limits of our compartment. We sleep, first singly, then by groups, and finally as a whole. Vainly we endeavour to ward off the coming lethargy. We set our thoughts on the sublime or beautiful, and try by an effort of will to hold them there. It is no use. Thought gradually slips away from its object, or the object glides out of the nerveless grasp of thought, and we are conquered by the heat. But what *is* heat that it should work such changes in moral and intellectual nature? Why should 'souls of fire' be the heirlooms of 'children of the sun'? Why are we unable to read 'Mill's Logic' or study the 'Kritik der Reinen Vernunft' with any profit in a Turkish bath? Heat, defined without reference to our sensations, is a peculiar kind of motion—motion, moreover, as strictly mechanical as the waves of the sea, or as the aërial vibrations which produce sound. The communication of this motion to the material atoms of the brain produces the moral and intellectual effects just referred to. Human action is only possible within a narrow zone of temperature. Transgress the limit on one side, and we are torpid by excess; transgress it on the other, and we are torpid by defect. The intellect is in some sense a function of temperature. Thus at 2 P.M. we wallowed in our cushions, drained of intellectual energy; six hours later, the stars were sown broadcast through space, and the mountains drew their outlines against the amber of the western sky. The mind was awake and active, and through her operations was shed that feeling of devotion which the mystery of creation ever inspires. Physically considered, however, the intellect of 2 P.M. differed from that of 8 P.M. simply in the amount of motion possessed by the molecules of the brain. You, my friend, know that it is not levity which prompts me to write thus. Matter, in relation to vital phenomena, has yet to be studied, and the command of Canute to the waves would be wisdom itself compared with any attempt to stop such inquiries. Let the tide rise, and let knowledge advance; the limits of the one are not more rigidly fixed than those of the other; and no worse infidelity could seize upon the mind than the belief that a man's earnest search after truth should culminate in his perdition. Fear not, my friend, but rest assured that as we understand *matter* better, *mind* will become capable of nobler and of wiser things.

The sun was high in heaven as we rolled from the station on the morning of the 2nd. I was in fair health, and therefore happy. The man who has work to do in the world, who loves his work, and joyfully invests his strength in the prosecution of it, needs but health to make him happy. Sooner or later every intellectual canker disappears before earnest work. Its influence, moreover, fills a wide margin beyond the time of its actual performance. Thus, to-day, I sang as I rolled along—not with boisterous glee, but with that serene and deep-lying gladness which becomes a man of my years and of my vocation. This happiness, however, had its roots in the past, and had I not been a worker previous to my release from London, I could not now have been so glad an idler. Nature, moreover, was in a pleasant mood; indeed, in any other country than Switzerland, the valley through which we sped would have produced excitement and delight. Noble fells, proudly grouped, flanked us right and left. Cloud-like woods of pines overspread them in broad patches, with between them spaces of the tenderest green; while here and there the rushing Rhine gleamed like an animating spirit amid the meadows.

Some philosophers inculcate an independence of external things, and a reliance upon the soul alone. But what would man be without Nature? A mere *capacity*, if such a thing be conceivable alone; potential, but not dynamic; an agent without an object. And yet how differently Nature affects different individuals! To one she is an irritant which evokes all the grandeur of the heart, while another is no more affected by her magnificence than are the beasts which perish. The one has halls and corridors within, in which to hang those images of splendour which Nature exhibits; the other has not even a châlet to offer for their reception. The countenance betrays, in some degree, the measure of endowment here. I know—*you* know—countenances, where the mind, shining through the eyes, conveys hints of inner bloom and verdure; of noble heights and deep secluded dells; of regions also unexplored and unexplorable, which in virtue of their mystery present a never-flagging charm to the mind. You, my friend, have experienced the feelings which an Alpine sunset wakes to life. You call it tender, but the tenderness resides in *you;* you speak of it as splendid, but the splendour is half your own. Creation sinks beyond the bottom of your eye, and finds its friend and interpreter in a region far behind the retina.

Hail to the Giants of the Oberland! there they stand, pyramid beyond pyramid, crest above crest. The zenith is blue, but the thick stratum of horizontal air invests the snowy peaks with a veil of translucent vapour, through which their vast and spectral outlines are clearly seen. As we roll on towards Thun this vapour thickens, while dense and rounded clouds burst heavenwards, as if let loose from a prison behind the mountains. The heavens darken, and the scowling atmosphere is cut by the lightning in sharp-bent bars of solid light. Afterwards comes the cannonade, and then the heavy rain-pellets which rattle with fury against the carriages. Again it clears, but not wholly. Stormy cumuli swoop round the mountains, between which, however, the illuminated ridges seem to swim in the transparent air.

At Thun I find my faithful and favourite guide, Johann Benen, of Laax, in the valley of the Rhone, the strongest limb and stoutest heart of my acquaintance in the Alps.[1]

---

[1] Benen's letter, in reply to mine, desiring to engage him, is, I think, worth inserting here.

HOCHGESCHÄTZTER HERR TYNDALL,—Indem ich mit Herrn Tugget (50) fünfzig Tage auf Reisen war und Heute erst nach Laax gekommen bin, habe ich Ihren werthen Brief von 22 Juni auch nur erst Heute erhalten; so dass ich denselben Ihnen auch nur Heute gleich beantworten konnte. Wo ich Ihnen mit Vergnügen melde,

We take the steamer to Interlaken, and while on the lake the heavens again darken, and the deck is flooded by the gushing rain. The dusky cloud-curtain is rent at intervals, and through the apertures thus formed gleams of sunlight escape, which draw themselves in parallel bars of extraordinary radiance across the lake. On reaching Interlaken, I drive to the steamer on the lake of Brientz, while my friend F. diverges to Grindelwald to seek a guide. We start at 6 P.M., with a purified atmosphere, and pass through scenes of serene beauty in the tranquil evening light. The bridge of Brientz has been carried away by the floods, the mail is intercepted, and I associate myself with a young Oxford man in a vehicle to Meyringen. The west wind has again filled the atmosphere with gloom, and after supper I spend an hour watching the lightning thrilling behind the clouds. The darkness is intense, and the intermittent glare correspondingly impressive. Now it is the east which is suddenly illuminated, now the west, now the heavens in front; now the visible light is evidently the fringe of an illuminated cloud which has caught the blaze of a discharge far down behind the mountains. Sometimes the lightning seems to burst, like a fireball, midway between the horizon and the zenith, spreading as a vast glory behind the clouds and revealing all their outlines. In front of me is a craggy summit, which indulges in intermittent shots of thunder; sharp, dry, and sudden, with scarcely an echo to soften them off.

dass ich immerhin bereit sein werde Sie zu begleiten wann und wohin Sie nur wünschen.

Herr Tyndall! Ich mache Ihnen meine Komplimente für das gute Zutrauen zu mir und hoffe noch an der Zeit gekommen zu sein um wieder Gelegen heit zu haben Sie bestens und baldigst zu bedienen.

Mit Hochschätzung und Emphelung,—Ihr Diener,

BENEN.

CHAPTER II

MEYRINGEN TO THE GRIMSEL, BY THE URBACHTHAL AND GAULI GLACIER

> 'Spring still makes spring in the mind
> When sixty years are told,
> Love wakes anew this throbbing heart,
> And we are never old.
> Over the winter glaciers
> I see the summer glow,
> And through the wild-piled snow drift
> The warm rose buds below.'

OUR bivouac at Meyringen was *le Sauvage*, who discharged his duty as a host with credit to himself and with satisfaction to us. F. has arrived, and in the afternoon of the 3rd we walk up the valley. Between Meyringen and Hof, the vale of Hasli is dammed across by a transverse ridge called the Kirchet, and the rocky barrier is at one place split through, forming a deep chasm with vertical sides through which the river Aar plunges. The chasm is called the Finsteraar-schlucht, and by the ready hypothesis of an earthquake its formation has been explained. Man longs for causes, and the weaker minds unable to restrain their hunger, often barter, for the most sorry theoretic pottage, the truth which patient inquiry would make their own. This proneness of the human mind to jump to conclusions, and thus shirk the labour of real investigation, is a most mischievous tendency. We complain of the contempt with which practical men regard theory, and, to confound them, triumphantly exhibit the speculative achievements of master minds. But the practical man, though puzzled, remains unconvinced; and why? Simply because nine out of ten of the theories with which he is acquainted are deserving of nothing better than contempt. Our master minds built their theoretic edifices upon the rock of fact; the quantity of fact necessary to enable them to divine the *law*, being a measure of individual genius, and not a test of philosophic system. As regards the Finsteraar-schlucht, instead of jumping to an earthquake to satisfy our appetite for 'deduction,' we must look at the circumstances. The valley of Hof lies above the mound of the Kirchet; how was this flat formed? Is it not composed of the sediment of a lake? Did not the Kirchet form the dam of this lake, a stream issuing from the latter and falling over the dam? And as the sea-waves find a weak point in the cliffs against which they dash, and gradually eat their way so as to form caverns with high vertical sides, as at the Land's End, a joint or fault or some other accidental weakness determining their line of action; so surely a mountain torrent rushing for ages over the Kirchet dam would be competent to cut itself a channel. The Kirchet itself has been moulded by the ancient glacier of the Aar. When Hof was a lake, that glacier had retreated, and from it issued the stream, the stoppage of which formed the lake. The stream finally cut itself a channel deep enough to drain the lake, and left the basis of green meadows as sediment behind; while through these meadows the stream that once overflowed their site now runs between grassy banks. Imagination is essential to the natural philosopher, but instead of indulging in off-hand theoretic guesses, he must regard the facts, discern their connection, and out of them reconstruct the world gone by.

Throughout the early part of this day the weather had been sulky, but towards evening the clouds were in many places torn asunder, revealing the blue of heaven and the direct beams of the sun. At midnight I quitted my bed to look at the weather, and found the sky spangled all over with stars. We were called at 4 A.M., an hour later than we intended, and the sight of the cloudless mountains was an inspiration to us all. At 5.30 A.M. we were off, crossing the valley of Hof, which was hugged round its margin by a light and silky mist. We ascended a spur which separated us from the Urbachthal, through which our route lay. The Aar for a time babbled in the distance, until, on turning a corner, its voice was suddenly extinguished by the louder music of the Urbach, rendered mellow and voluminous by the resonance of the chasm into which the torrent leaped. The sun was already strong, and the world on which he shone was grand and beautiful. His yellow light glimmered from the fresh green leaves; he smote with glory the boles and the plumes of the pines; soft shadows fell from shrub and rock on the emerald pastures; snow-peaks were in sight, cliffy summits also, without snow or verdure, but in many cases, buttressed by slopes of soil which bore a shaggy growth of trees. The grass over which we passed was sown with orient pearls; to the right of us rose the bare cliffs of the Engelhörner, broken at the top into claw-shaped masses which were turned, as if in spite, against the serene heaven. Benen walked on in front, a mass of organised force, silent, but emitting at times a whistle which sounded like the piping of a lost chamois. Hark to an avalanche! In a hollow of the Engelhörner a mass of snow had found a lodgment; melted by the warm rock, its foundation was sapped, and down it came in a thundering cascade. The thick pinewoods to our right were furrowed by the tracks of these destroyers, the very wind of which, it is affirmed, tears up distant trees by the roots.

For a time our route lies through a spacious valley which now turns to the left, narrows to a gorge, and winds away amid the mountains. Along its bottom the hissing river rushes; this we cross, climb the wall of a *cul de sac*, and from its rim enjoy a glorious view. The Urbachthal has been the scene of vast glacier action; with tremendous energy the ice of other days must have been driven by its own gravity through the narrow gorge, planing and fluting and scoring the rocks. Looking at these charactered crags, one's thoughts involuntarily revert to the ancient days, and from a few scattered observations we restore in idea a state of things which had disappeared from the world before the development of man. Whence this wondrous power of reconstruction? Who will take the step which shall connect the faculties of the human mind with the physics of the human brain? Was this power locked like latent heat in ancient inorganic nature, and developed as the ages rolled? Are other and grander powers still latent in nature, which shall come to blossom in another age? Let us question fearlessly, but, having done so, let us avow frankly that at bottom we know nothing; that we are embedded in a mystery towards the solution of which no whisper has been yet conceded to the listening intellect of man.

The world of life and beauty is now retreating, and the world of death and beauty is at hand. We are soon at the end of the Gauli Glacier, from which our impetuous friend the Urbach rushes, and turn into a châlet for a draught of milk. The Senner within proved an extortioner —'*ein unverschämter Hund;*' but let him pass without casting a speck upon the brightness of our enjoyment. We work along the flank of the glacier to a point which commands a view of the cliffy barrier which it is the main object of our journey to pass. From a range of snow-peaks linked together by ridges of black rock, the Gauli Glacier falls, at first steeply as snow, then more gently as ice. We scan the mountain barrier to ascertain where it ought to be attacked. No one of us has ever been here before, and the scanty scraps of information which we have received tell us that at one place only is the barrier passable. We may reach the summit at several points from this side, but all save one, we are informed, lead to the brink of intractable precipices which fall sheer to the Lauteraar Glacier. We observe, discuss, and finally decide upon a point of attack. We enter upon the glacier; black chasms yawn here and there through the superincumbent snow, but there is no real difficulty. We cross the glacier and reach the opposite slopes; our way first lies up a moraine, and afterwards through the snow; a laborious ascent brings us close to the ridge, and here we pause once more in consultation. There is a gentle indentation to our left, and a cleft in the rocks to our right; our information points decidedly to the latter, but still our attention is attracted by the former. 'Shall we try the saddle, sir? I think we shall get down,' asks Benen. 'I think so too; let us make for it,' is my reply.

The winter snows were here thickly laid against the precipitous crags; the lower part of the buttress thus formed had broken away from the upper, which still clung to the rocks, and the whole ridge was thus defended by a profound chasm, called in Switzerland a bergschrund. At some places portions of snow had fallen away from the upper slope and partially choked the schrund, closing, however, its mouth only, and on this snow we were now to seek a footing. Benen and myself were loose coming up, F. and his guide were tied together; but now F. declares that we must all be attached, as it would injure his stomach to see us try the schrund singly. We accordingly rope ourselves, and advance along the edge of the fissure to one of the places where it is partially stopped. At this place a vertical wall of snow faces us. Our leader carefully treads down the covering of the chasm; and having thus rendered it sufficiently rigid to stand upon, he cuts a deep gap with his ice-axe in the opposing wall. Into the gap he tries to force himself, but the mass yields, and he falls back, sinking deeply in the snow of the schrund. You must bear in mind that he stands right over the fissure, which is merely bridged by the snow. I call out 'Take care!' he responds 'All right!' and returns to the charge. He hews a deeper and more ample gap; strikes his axe into the slope above him, and leaves it there; buries his hands in the yielding mass, and raising his body on his two arms, as on a pair of pillars, he lifts himself into the gap. He is thus clear of the schrund, and soon anchors his limbs in the snow above. I am speedily at his side, and we both tighten the rope as our friend F. advances. With perfect courage and a faultless head, he has but one disadvantage, and that is an excess of weight of at least two stone. In his first attempt the snow ledge breaks, and he falls back; but two men are now at the rope, the tension of which, aided by his own activity, prevents him from sinking far. By a second effort he clears the difficulty, is followed by his guide, and all four of us are now upon the slope above the chasm. Had you, unaccustomed to mountain climbing, found yourself upon such an incline, you would have deemed it odd. Its steepness was greater than that of a cathedral roof, while below us, and within a few yards of us, the slope was cut by a chasm into which it would be certain death to fall. Education enables us to regard a position of this kind almost with indifference, still the work was by no means unexciting. In this early stage of our summer performances, it required perfect trust in our leader to keep our minds at ease. A doubt of him would have introduced moral and physical weakness amongst us all; but the feebleness of uncertainty was unfelt; we made use of all our strength, and consequently succeeded with comparative ease. We are now near the top of the saddle, separated from it, however, by a very steep piece of snow; this is soon overcome, and a cheer at the summit announces that our escape is secured.

The indentation, in fact, forms the top of a kind of chimney or cut in the rocks, which leads right down to the Lauteraar Glacier. It is steep, but we know that it is feasible, and we pause contentedly upon the summit to scan the world of mountains extending beyond. The Schreckhorn particularly interests my friend F. It had been tried in successive years by Mr. A. without success, and now F. had set his heart on climbing it. The hope of doing so from this side is instantly extinguished, the precipices are so smooth and steep. Elated with our present success, I release myself from the rope and spring down the chimney, preventing the descent from quickening to an absolute fall by seizing at intervals the projecting rocks. Once an effort of this kind shakes the alpenstock from my hand; it slides along the débris, reaches a snow slope, shoots down it, and is caught on some shingle at the bottom of the slope. Benen wishes to get it for me, but I am instantly after it myself. Quickly skirting the snow, which, without a staff, cannot be trusted, an *arête* is reached, from which a jump lands me on the débris: it yields and carries me down; passing the alpenstock I seize it, and in an instant am master of all my motions. Another snow slope is reached, down which I shoot to the rocks at the bottom, and there await the arrival of Benen. He joins me immediately; F. and his guide, however, choosing a slower mode of descent. We have

Meyringen to the Grimsel

diverged from the deep cut of the chimney, rough rocks are in our way; to these Benen adheres while I, hoping to make an easier descent through the funnel itself, resort to it. It is partially filled with indurated snow, but underneath this a stream rushes, and my ignorance of the thickness of the roof renders caution necessary. At one place the snow is broken quite across, and a dark tunnel, through which the stream rushes, opens immediately below me. My descent is thus cut off, and I cross the couloir to the opposite rocks, climb them, and find myself upon the summit of a ledged precipice, below which Benen halts, and watches me as I descend it. On one of the ledges my foot slips; a most melancholy whine issues from my guide, as he suddenly moves towards me to render what help he can; but the slip in no way compromises the firmness of my grasp; I reach the next ledge, and in a moment am clear of the difficulty. We drop down the mountain together, quit the rocks, and reach the ice of the glacier, where we are soon joined by F. and his companion. Turning round now we espy a herd of seven chamois on one of the distant slopes of snow. The telescope reduces them to five full-grown animals and two pretty little kids, fair and tender tenants of so wild a habitation. Down we go along the glacier, with the sun on our backs, his beams streaming more and more obliquely against the ice. The deeper glacier pools are shaded in part by their icy banks, and through the shadowed water needles of ice are already darting: all day long the molecules had been kept asunder by the antagonistic heat; their enemy is now withdrawn, and they lock themselves together in a crystalline embrace. Through a reach of merciless shingle, which covers the lower part of the glacier, we now work our way; over green pastures; over rounded rocks; up to the Grimsel Hotel, which, uncomfortable as it is, is reached with pleasure by us all.

CHAPTER III

THE GRIMSEL AND THE ÆGGISCHHORN

> 'Thou trowest
> How the chemic eddies play
> Pole to pole, and what they say;
> And that these grey crags
> Not on crags are hung,
> But beads are of a rosary
> On prayer and music strung.'

GRANDLY on the morning of the 5th, the sun rose over the mountains, filling earth and air with the glory of his light. This Grimsel is a weird region—a monument carved with hieroglyphics more ancient and more grand than those of Nineveh or the Nile. It is a world disinterred by the sun from a sepulchre of ice. All around are evidences of the existence and might of the glacier which once held possession of the place. All around the rocks are carved, and fluted, and polished, and scored. Here and there angular pieces of quartz, held fast by the ice, inserted their edges into the rocks and scratched them like diamonds, the scratches varying in depth and width according to the magnitude of the cutting stone. Larger masses, held similarly captive, scooped longitudinal depressions in the rocks over which they passed, while in many cases the polishing must have been effected by the ice itself. A raindrop will wear a stone away, much more would an ice surface, squeezed into perfect contact by enormous pressure, rub away the asperities of the rocks over which for ages it was forced to slide. The rocks thus polished by the ice itself are exceedingly smooth, and so slippery that it is impossible to stand on them where their inclination is at all considerable. But what a world it must have been when the valleys were thus filled! We can restore the state of things in thought, and in doing so we submerge many a mass which now lifts its pinnacle skyward. Switzerland in

The Grimsel and Æggischhorn

those days could not be so grand as it is now. Pour ice into those valleys till they are filled, and you eliminate those contrasts of height and depth on which the grandeur of Alpine scenery depends. Instead of skyey pinnacles and deep-cut gorges we should have an icy sea dotted with dreary islands formed by the highest mountain tops.

In the afternoon I strolled up to the Siedelhorn; a mountain often climbed by tourists for the sake of the prospect it commands. This is truly fine. As I stood upon the broken summit of the mountain the sun was without a cloud; and his rays fell directly against the crown and slopes of the Galenstock at the base of which lay the glacier of the Rhone. The level sea of *névé* above the great ice-cascade, the fall itself, and the terminal glacier below the fall were all apparently at hand. At the base of the fall the ice, as you know, undergoes an extraordinary transformation; it reaches this place more or less amorphous, it quits it most beautifully laminated, the change being due to the pressure endured by the ice at the bottom of the fall. The wrinkling of the glacier here was quite visible, the dwindling of the wrinkles into bands, and the subdivision of these bands into lines which mark the edges of the laminæ of which the glacier at this place is composed. Beyond, amid the mountains at the opposite side of the Rhone valley, lies the Gries Glacier, half its snow in shadow, and half illuminated by the sinking sun. Round farther to the right stand the Monte Leone and other grand masses, the grandest here being the Mischabel with its crowd of snowy cones. Jumping a gap in the mountains, we hit the stupendous cone of the Weisshorn, which slopes to meet the inclines of the Mischabel, and in the wedge of space carved out between the two, the Matterhorn lifts its terrible head. Wheeling farther in the same direction, we at length strike the mighty spurs of the Finsteraarhorn, between two of which lies the Oberaar Glacier. Here is no turmoil of crevasses, no fantastic ice-pinnacles, nothing to indicate the operation of those tremendous forces by which a glacier sometimes rends its own breast, but soft and quiet it reposes under its shining coverlet of snow. The grimmest fiend of the Oberland closes the view at the head of the Lauteraar Glacier; this is the Schreckhorn, whose cliffs on this side no mountaineer will ever scale. Between the Schreckhorn and Finsteraarhorn a curious group of peaks encircle a flat snow-field, from which the sunbeams are flung in blazing lines. Immediately below is the Unteraar Glacier, with a long black streak upon its back, bent hither and thither, like a serpent in the act of wriggling down the valley. Beyond it and flanking it, is a range of mountains with a crest of vertical rock, hacked into indentations which suggest a resemblance to a cockscomb; and to the very root of the comb the mountains have been cut away by the ancient ice. A scene of unspeakable desolation it must have been when Europe was thus encased in frozen armour, and when even the showers of her western isles fell solid from the skies,—when glaciers teemed from the shoulders of Snowdon and Scawfell, and when Llanberis and Borrodale were ploughed by frozen shares,—when the Reeks of Magillicuddy sent down giant navigators to delve out space for the Killarney lakes, and to saw through the mountains the Gap of Dunloe. Evening comes, and we move downwards, down amid heaped boulders; down over the tufted alp; down with headlong speed over the *roches moutonnés* of the Grimsel Pass, making long springs at intervals, over the polished inclines, and reaching the hospice as its bell rings its hungry inmates to their evening meal.

F. and I had arranged to pay a visit to the Schreckhorn on the following day. He was not well, and wisely stayed at home; I was not well, and unwisely went. The day was burning hot, and the stretch of glacier from the Grimsel to the Strahleck very trying. We, however, gained the summit of the pass, and from it scanned the peak which F. wished to assail. An adjacent peak had been surmounted by M. Desor and some friends, at the time of Agassiz's observations on the Lower Aar Glacier, but the summit they attained was about eighty feet below the true one, and to pass from one to the other they found impossible. We concluded that the ascent, though difficult, might be accomplished by spending the previous night upon the Strahleck.[1] I had my heart on other summits, and was unwilling to divert from them the time and trouble which the Schreckhorn would demand. Neither could I advise F. to try it, as his power among rocks like those of the Schreckhorn was still to be tested. The idea of climbing this pinnacle was therefore relinquished by us both.

On Saturday, accompanied by my friend and former fellow climber H., I ascended from Viesch to the Hotel Jungfrau on the slope of the Æggischhorn, and in the evening of the same day walked up to the summit of the mountain alone. As is usual with me, I wandered unconsciously from the beaten track, and had to make my way amid the chaos of crags which Nature, in her ruinous moods, had shaken from the mountain. From these I escaped to a couloir, filled in part with loose débris, and down which the liberated boulders roll. My ascent was quick, and I soon found myself upon the crest of broken rocks which caps the mountain. This peak and those adjacent, which are similarly shattered, exhibit a striking picture of the ruin which Nature inflicts upon her own creations. She buildeth up and taketh down. She lifts the mountains by her subterranean energies, and then blasts them by her lightnings and her frost. Thus grandly she rushes along the 'grooves of change' to her unattainable repose. Is it unattainable? The incessant tendency of material forces is toward final equilibrium; and if the quantity of this tendency be finite, a time of repose must come at last. If one portion of the universe be hotter than another, a flux instantly sets in to equalise the temperatures; while winds blow and rivers roll in search of a stable equilibrium. Matter longs for rest; when is this longing to be fully satisfied? If satisfied, what then? The state to which material nature tends is not one of perfection, but of death. Life is only compatible with mutation; and when the attractions and

[1] It was actually accomplished from this side, a few days after my visit, by the Rev. Leslie Stephen.

Mountaineering in 1861

repulsions of material atoms have been satisfied to the uttermost, life ceases, and the world thenceforward is locked in everlasting sleep.

A wooden cross bleached by many storms surmounts the pinnacle of the Æggischhorn, and at the base of it I now take my place and scan the surrounding scene. Down from its birthplace in the mountains comes that noblest of ice-streams, the Great Aletsch Glacier. Its arms are thrown round the shoulders of the Jungfrau, while from the Monk and the Trugberg, the Gletscherhorn, the Breithorn, the Aletschhorn, and many another noble pile, the tributary snows descend and thicken into ice. The mountains are well protected by their wintry coats, and hence the quantity of débris upon the glacier is comparatively small; still, along it we notice dark longitudinal streaks, which occupy the position the moraines would assume had matter sufficient to form them been cast down. Right and left from these longitudinal bands finer curves sweep across the glacier, twisted here and there into complex windings. They mark the direction in which the subjacent ice is laminated. The glacier lies in a curved valley, the side towards which its convex curvature is turned, is thrown into a state of strain, the ice breaks across the line of tension, and a curious system of oblique glacier ravines is thus produced. From the snow-line which crosses the glacier above the Faulberg a pure snow-field stretches upward to the Col de la Jungfrau; the Col which unites the maiden to her sacerdotal neighbour. Skies and summits are to-day without a cloud, and no mist or turbidity interferes with the sharpness of the outlines. Jungfrau, Monk, Eiger, Trugberg, cliffy Strahlgrat, stately, lady-like Aletschhorn, all grandly pierce the empyrean. Like a Saul of mountains the Finsteraarhorn overtops all his neighbours; then we have the Oberaarhorn, with the riven glacier of Viesch rolling from his shoulders. Below is the Mârjelin See, with its crystal precipices and its floating icebergs, snowy white, sailing on a blue-green sea. Beyond is the range which divides the Valais from Italy. Sweeping round, the vision meets an aggregate of peaks which look, as fledglings to their mother, towards the mighty Dom. Then come the repellent crags of Mont Cervin; the idea of moral savagery, of wild untamable ferocity, mingling involuntarily with our contemplation of the gloomy pile. Next comes an object, scarcely less grand, conveying it may be even a deeper impression of majesty and might, than the Matterhorn itself,—the Weisshorn, perhaps the most splendid object in the Alps. But beauty is associated with its force, and we think of it, not as cruel, but as grand and strong. Farther to the right the Great Combin lifts up his bare head; other peaks crowd around him; while at the extremity of the curve round which our gaze has swept rises the sovran crown of Mont Blanc. And now, as the day sinks, scrolls of pearly clouds draw themselves around the mountain crests, being wafted from them into the distant air. They are without colour of any kind; still, by grace of form, and as the embodiment of lustrous light and most tender shade, their beauty is not to be described.

CHAPTER IV

THE BEL ALP

'Happy, I said, whose home is here;
Fair fortunes to the Mountaineer.'

Up to Tuesday the 13th I remained at the Æggischhorn, sauntering over the Alps, or watching dreamily the mutations of light and shade upon the mountains. On this day I accompanied a party of friends to the Mârjelin See, skirted the lake, struck in upon the glacier, and having heard much of the position and the comfort of a new hotel upon the Bel Alp, I resolved to descend the glacier and pay the place a visit. The Valais range had been already clouded before we quitted the hotel, still the sun rode unimpeded in the higher heavens. Vast vapour masses, however, continued to thrust themselves forth like arms into the upper air; spreading laterally, they became entangled with each other, and thus the mesh of cloud became more continuous and obscure. Having tried in vain to daunt an English maiden whom I led among the crevasses, I separated from my companions, who had merely made an excursion from the hotel, and my friend T., Benen, and myself commenced the descent of the glacier. The clouds unlocked themselves, thunder rung and echoed amid the crags of the Strahlgrat, accompanied by a furious downpour of rain. We crouched for a time behind a parapet of ice until the rain seemed to lighten, when we emerged, and went swiftly down the glacier. Sometimes my guide was in advance among the ice-hills and ravines, sometimes myself, an accident now and then giving the one an advantage over the other. The rain again commencing, we escaped from the ice to the flanking hillside, and hid ourselves for a time under a ledge of rock; being finally washed out of our retreat by the rush of water.

The Bel Alp

The rain again lightens, and we are off. The glacier is here cut up into oblique valleys of ice, these being subdivided by sharp-edged crevasses. We advance swiftly along the ridges which divide vale from vale, but these finally abut against the mountain, and we are compelled to cross from ridge to ridge. T. follows Benen, and I trust to my own devices. Joyously we strike our axes into the crumbling crests, and make our way rapidly between the chasms. The sunshine gushes down upon us for a time, and partially dries our drenched clothes, after which the atmosphere again darkens. A storm is brewing, and we urge ourselves to a swifter pace. At some distance to our left, we observe upon the ice a group of persons, consisting of two men, a boy, and an old woman. They are engaged beside a crevasse, and a thrill of horror shoots through me, at the thought of a man being possibly between its jaws. We quickly join them, and find an unfortunate cow firmly jammed between the frozen sides of the fissure, groaning most piteously, but wholly unable to move. The men had possessed themselves of a bad rope and a common hatchet, and were doing their utmost to rescue the animal; but their means were inadequate, and their efforts ill-directed. They had passed their rope under the animal's tail, hoping thereby to raise its heavy haunches from the chasm; of course the noose slipped along the tail and was utterly useless. 'Give the brute space, cut away the ice which presses the ribs, and *you* step upon that block which stops the chasm, and apply your shoulders to the creature's buttocks.' The ice splinters fly aloft, under the vigorous strokes of Benen. T. suggests that one rope should be passed round the horns, so as to enable all hands to join in the pull. This is done. 'Pass your rope between the animal's hind legs instead of under its tail.' This is also done. Benen has loosened the ice which held the ribs in bondage, and now like mariners heaving an anchor, we all join in a tug, timing our efforts by an appropriate exclamation. The brute moves, but extremely little; again the cry, and again the heave—she moves a little more. This is repeated several times till her fore-legs are extricated and she throws them forward on the ice. We now apply our efforts to her hinder parts, and succeed in placing the animal upon the glacier, panting and trembling in all her fibres. 'Fold your rope, Johann, and onward; the day is darkening, and we know not what glacier work is still before us.' On we went. Hark once more, to the thunder, now preceded by vivid lightning gleams which flash into my eyes from the polished surface of my axe. Gleam follows gleam, and peal succeeds peal with terrific grandeur; and the loaded clouds send down from all their fringes dusky streamers of rain. These look like waterspouts, so dense is their texture. Furious as was the descending shower; hard as we were hit by the mixed pellets of ice and water, I scarcely ever enjoyed a scene more. Grandly the cloud-besom swept the mountains, their colossal outlines looming at intervals like overpowered Titans struggling against their doom.

We are now entangled in crevasses, the glacier is impracticable, and we are forced to retreat to its western shore. We pass along the lateral moraine: rough work it proved, and tried poor T. severely. The mountain slope to our left becomes partially clothed with pines, but such spectral trees! Down the glacier valley wild storms had rushed, stripping the trunks of their branches, and the branches of their leaves, and leaving the treewrecks behind, as if spirit-stricken and accursed. We pause and scan the glacier, and decide upon a place to cross it. Our home is in sight, perched upon the summit of the opposite mountain. On to the ice once more, and swiftly over the ridges towards our destination. We reach the opposite side, wet and thirsty, and face the steep slope of the mountain; slowly we ascend it, strike upon a beaten track, and pursuing it, finally reach the pleasant auberge at which our day's journey ends.

If you and I should be ever in the Alps together, I shall be your guide from the Æggischhorn to the Bel Alp. You shall choose yourself whether the passage is to be made along the glacier, as we made it, or along the grassy mountain side to the Rieder Alp, and thence across the glacier to our hotel. Here, if the weather smile upon us,

The Bel Alp

we may halt for two or three days. From the hotel on the Æggischhorn slope an hour and a half's ascent is required to place the magnificent view from the summit of the Æggischhorn in your possession. But from the windows of the hotel upon the Bel Alp noble views are commanded, and you may sit upon the bilberry slopes adjacent, in the presence of some of the noblest of the Alps. And if you like wildness, I will take you down to the gorge in which the Aletsch Glacier ends, and there chill you with fear. I went down to this gorge on the 14th, and shrunk from the edge of it at first. A pine-tree stood sheer over it; bending its trunk at a right angle near its root, it laid hold of a rock, and thus supported itself above the chasm. I stood upon the horizontal part of the tree, and, hugging its upright stem, looked down into the gorge. It required several minutes to chase away the timidity with which I hung over this savage ravine; and, as the wind blew more forcibly against me, I clung with more desperate energy to the tree. In this wild spot, and alone, I watched the dying fires of the day, until the latest glow had vanished from the mountains.

And if you like to climb for the sake of a wider horizon, you shall have your wish at the Bel Alp. High above it towers a grey pinnacle called the Sparrenhorn, and two hours of moderate exertion will place you and me together upon that point. I went up there on the 15th. To the observer from the hotel the Sparrenhorn appears as an isolated peak; it forms, however, the lofty end of a narrow ridge, which is torn into ruins by the weather, flanking on the east the forsaken bed of a *névé*, and bounding on the west the Ober Aletsch Glacier. In front of me was a rocky promontory like the Abschwung, right and left of which descended two streams of ice, which welded themselves to a common trunk. This glacier scene was perfectly unexpected and strikingly beautiful. Nowhere have I seen such perfect repose, nowhere more tender curves or finer structural lines, forming loops across the glacier. The stripes of the moraine bending along its surface contribute to its beauty, and its deep seclusion gives it a peculiar charm. It is a river so protected by its bounding mountains that no storm can ever reach it, and no billow disturb the perfect serenity of its rest. The sweep of the Aletsch Glacier is also mighty as viewed from this point, and from no other could the Valais range seem more majestic. It is needless to say a word about the grandeur of the Dom, the Cervin, and the Weisshorn, all of which, and a great deal more, are commanded from this point of view. Surely you and I must clamber thither, and if your feet refuse their aid I will pass my strap around your waist and draw you to the top.

CHAPTER V

REFLECTIONS

'The world was made in order,
And the atoms march in tune.'

THE aspects of Nature are more varied and impressive in Alpine regions than elsewhere. The mountains themselves are permanent objects of grandeur. The effects of sunrise and sunset; the formation and distribution of clouds; the discharge of electricity, such as we witnessed a day or two ago; the precipitation of rain, hail, and snow; the creeping of glaciers and the rushing of rivers; the colouring of the atmosphere and its grosser action in the case of storms;—all these things tend to excite the feelings and to bewilder the mind. In this entanglement of phenomena it seems hopeless to seek for law or orderly connection. And before the thought of law dawned upon the human mind men naturally referred these inexplicable effects to personal agency. The savage saw in the fall of a cataract the leap of a spirit, and the echoed thunder-peal was to him the hammer-clang of an exasperated god. Propitiation of these terrible powers was the consequence, and sacrifice was offered to the demons of earth and air.

But the effect of time appears to be to chasten the emotions and to modify the creations which depend upon them alone, by giving more and more predominance to the intellectual power of man. One by one natural phenomena were associated with their proximate causes; this process still continues, and the idea of direct personal volition mixing itself in the economy of nature is retreating more and more. Many of us fear this tendency; our faith and feelings are dear to us, and we look with suspicion and dislike on any philosophy which would deprive us of the relations in which we have been accustomed to believe, as tending directly to dry up the soul. Probably every change from ancient savagery to our present enlightenment excited, in a greater or less degree, a fear of this kind. But the fact is, that we have not at all determined whether the form under which they now appear in the world is necessary to the prosperity of faith and feeling. We may err in linking the imperishable with the transitory, and confound the living plant with the decaying pole to which it clings. My object, however, at present is not to argue, but to mark a tendency. We have ceased to propitiate the powers of Nature,—ceased even to pray for things in *manifest* contradiction to natural laws. In Protestant countries, at least, I think it is conceded that the age of miracles is past.

The general question of miracles is at present in able and accomplished hands; and were it not so, my polemical acquirements are so limited, that I should not presume to enter upon a discussion of this subject on its entire merits. But there is one little outlying point, which attaches itself to the question of miracles, and on which a student of science may, without quitting the ground which strictly belongs to him, make a few observations. If I should err here, there are many religious men in this country quite competent to correct me; and did I not feel it to be needless, I should invite them to do so. I shall, as far as possible, shut out in my brief remarks the exercise of mere *opinion*, so that if I am wrong, my error may be immediately reduced to demonstration.

At the auberge near the foot of the Rhone Glacier, I met in the summer of 1858, an athletic young priest, who, after he had accomplished a solid breakfast and a bottle of wine, informed me that he had come up to 'bless the mountains,' this being the annual custom of the place. Year by year the Highest was entreated to make such meteorological arrangements as should ensure food and shelter for the flocks and herds of the Valaisians. A diversion of the Rhone, or a deepening of the river's bed, would have been of incalculable benefit to the inhabitants of the valley at the time I mention. But the priest would have shrunk from the idea of asking the Omnipotent to open a new channel for the river, or to cause a portion of it to flow up the Mayenwand, over the Grimsel Pass, and down the vale of Oberhasli to Brientz. This he would have deemed *a miracle*, and he did not come to ask the Creator to perform miracles, but to do something which he manifestly thought lay quite within the bounds of the natural and non-miraculous. A Protestant gentleman who was present at the time, smiled at this recital. He had no faith in the priest's blessing, still he deemed the prayer actually offered to be different in kind from a request to open a new river-cut, or to cause the water to flow up-hill.

In a similar manner we all smile at the poor Tyrolese priest, who, when he feared the bursting of a glacier, offered the sacrifice of the mass upon the ice as a means of averting the calamity. That poor man did not expect to convert the ice into adamant, or to strengthen its texture so as to enable it to withstand the pressure of the water; nor did he expect that his sacrifice would cause the stream to roll back upon its source and relieve him, by a miracle, of its presence. But beyond the boundaries of his knowledge lay a region where rain was generated he knew not how. He was not so presumptuous as to expect a miracle, but he firmly believed that in yonder cloud-land matters could be so arranged, without trespass on the miraculous, that the stream which threatened him and his flock should be caused to shrink within its proper bounds.

Both the priests fashioned that which they did not understand to their respective wants and wishes; the unintelligible is the domain of the imagination. A similar state of mind has been prevalent among mechanicians; many of whom, and some of them extremely skilful ones, were occupied a century ago with the question of a *perpetual motion*. They aimed at constructing a machine which should execute work without the expenditure of power; and many of them went mad in the pursuit of this object. The faith in such a consummation, involving as it did immense personal interest to the inventor, was extremely exciting, and every attempt to destroy this faith was met by bitter resentment on the part of those who held it. Gradually, however, the pleasant dream dissolved, as men became more and more acquainted with the true functions of machinery. The hope of getting work out of mere mechanical combinations, without expending power, disappeared; but still there remained for the mechanical speculator a cloud-land denser than that which filled the imagination of the Tyrolese priest, and out of which he still hoped to evolve perpetual motion. There was the mystic store of chemic force, which nobody understood; there were heat and light, electricity and magnetism, all competent to produce mechanical motions.[1] Here, then, is the mine in which we must seek our gem. A modified and more refined form of the ancient faith revived; and, for aught I know, a remnant of sanguine designers may at the present moment be engaged on the problem which like-minded men in former years left unsolved.

And why should a perpetual motion, even under modern conditions, be impossible? The answer to this question is the statement of that great generalisation of modern science, which is known under the name of the *Conservation of Force*. This principle asserts that no power can make its appearance in Nature without an equivalent expenditure of some other form of power; that natural agents are so related to each other as to be mutually convertible, but that no new agency is created. Light runs into heat; heat into electricity; electricity into magnetism; magnetism into mechanical force; and mechanical force again into light and heat. The Proteus changes, but he is ever the same; and his changes in Nature, supposing no miracle to supervene, are the expression, not of spontaneity, but of *physical necessity*. One primal essence underlies all natural phenomena—and that is MOTION. Every aspect of Nature is a quality of motion. The atmosphere is such by its power of atomic motion. The glacier resolves itself to water, the water to transparent vapour, and the vapour to untrans-

---

[1] See Helmholtz—'Wechselwirkung der Naturkräfte.'

Reflections

parent cloud, by changes of motion. The very hand which moves this pen involves in its mechanical oscillation over this page the destruction of an equivalent amount of motion of another kind. A perpetual motion, then, is deemed impossible, because it demands the creation of force, whereas the principle of Nature is, no creation but infinite conversion.

It is an old remark that the law which moulds a tear also rounds a planet. In the application of law in Nature the terms great and small are unknown. Thus the principle referred to teaches us that the south wind gliding over the crest of the Matterhorn is as firmly ruled as the earth in its orbital revolution round the sun; and that the fall of its vapour into clouds is exactly as much a matter of necessity as the return of the seasons. The dispersion, therefore, of the slightest mist by the special volition of the Eternal, would be as much a miracle as the rolling of the Rhone up the precipices of the Mayenwand. It seems to me quite beyond the present power of science to demonstrate that the Tyrolese priest, or his colleague of the Rhone valley, asked for an impossibility in praying for good weather; but science *can* demonstrate the incompleteness of the knowledge of Nature which limited their prayers to this narrow ground; and she may lessen the number of instances in which we 'ask amiss,' by showing that we sometimes pray for the performance of a miracle when we do not intend it. She does assert, for example, that without a disturbance of natural law, quite as serious as the stoppage of an eclipse, or the rolling of the St. Lawrence up the Falls of Niagara, no act of humiliation, individual or national, could call one shower from heaven, or deflect towards us a single beam of the sun. Those, therefore, who believe that the miraculous is still active in nature, may, with perfect consistency, join in our periodic prayers for fair weather and for rain: while those who hold that the age of miracles is past, will refuse to join in such petitions. And they are more especially justified in this refusal by the fact that the latest conclusions of science are in perfect accordance with the doctrine of the Master Himself, which manifestly was that the distribution of natural phenomena is not affected by moral or religious causes. 'He maketh His sun to rise on the evil and on the good, and sendeth rain on the just and on the unjust.' Granting 'the power of Free Will in man,' so strongly claimed in his admirable essay by the last defender of the belief in miracles,[1] and assuming the efficacy of free prayer to produce changes in external nature, it necessarily follows that natural laws are more or less at the mercy of man's volition, and no conclusion founded on the assumed permanence of those laws would be worthy of confidence.

These considerations have been already practically acted upon by individual ministers of the Church of England; and it is one of the most cheering signs of the times to see such men coming forward to prepare the public mind for changes, which though inevitable, could hardly, without due preparation, be wrought without violence. Iron is strong; still, water in crystallising will shiver an iron envelope, and the more rigid the metal is, the worse for its safety. There are men of iron among us who would encompass human speculation by a rigid envelope, hoping thereby to chain the energy, but in reality dooming what they wish to preserve to more certain destruction. If we want an illustration of this we have only to look at modern Rome. In England, thanks to men of the stamp to which I have alluded, scope is gradually given to thought for changes of aggregation, and the envelope slowly alters its form in accordance with the necessities of the time.

[1] Professor Mansel.

CHAPTER VI

ASCENT OF THE WEISSHORN

> ' In his own loom's garment drest,
> By his proper bounty blest,
> Fast abides this constant giver,
> Pouring many a cheerful river,
> To far eyes an aërial isle,
> Unploughed, which finer spirits pile;
> Which morn, and crimson evening, paint
> For bard, for lover, and for saint;
> The country's core,
> Inspirer, prophet, evermore!'

ON Friday the 16th of August I rose at 4.30 A.M.; the eastern heaven was hot with the glow of the rising sun, and against the burning sky the mountain outlines were most impressively drawn. At 5.30 I bade goodbye to the excellent little auberge, and engaging a porter to carry my knapsack, went straight down the mountain towards Briegg. Beyond the end of the present ice the land gives evidence of vast glacier operations. It is scooped into hollows and raised into mounds; long ridges, sharpening to edges at the top, indicating the stranded moraines of the ancient glacier. And these hollows, and these hills, over which the ice had passed, destroying every trace of life which could possibly find a lodgment in them, were now clothed with the richest verdure. And not to vegetable life alone did they give support, for a million grylli chirruped in the grass. Rich, sapid meadows spread their emerald carpets in the sun; nut trees and fruit trees glimmered as the light fell upon their quivering leaves. Thus sanative nature healed the scars which she had herself inflicted. The road is very rough a part of the way to Briegg; let us trust that before your arrival it will be improved. I took the diligence to Visp, and engaged a porter immediately to Randa. I had sent Benen thither, on reaching the Bel Alp, to seek out a resting-place whence the Weisshorn might be assailed. On my arrival I learned that he had made the necessary reconnaissance, and entertained hopes of our being able to gain the top.

This noble mountain had been tried on various occasions and from different sides by brave and competent men, but had never been scaled; and from the entries in the travellers' books I might infer that formidable obstacles stood in the way of a successful ascent. The peak of the mountain is not visible at Randa, being far withdrawn behind the Alps. Beyond the Biezbach its ramparts consist of a craggy slope crowned above by three tiers of rocky strata. In front of the hotel is a mountain slope with pines clinging to its ledges, while stretching across the couloir of the Biezbach the divided ramparts are connected by battlements of ice. A quantity of débris which has been carried down the couloir spreads out in the shape of a fan at the bottom; near the edge of this débris stands a group of dingy houses, and close alongside them our pathway up the mountain runs.

Previous to quitting Randa I had two pair of rugs sewed together so as to form two sacks. These and other coverlets intended for my men, together with our wine and provisions, were sent on in advance of us. At 1 P.M., on the 18th of August, we, that is Benen, Wenger, and myself, quitted the hotel, and were soon zigzagging among the pines of the opposite mountain. Wenger had been the guide of my friend F., and had shown himself so active and handy on the Strahleck, that I commissioned Benen to engage him. During the previous night I had been very unwell, but I hoped that the strength left me, if properly applied, and drained to the uttermost, would still enable me to keep up with my companions. As I climbed the slope I suffered from intense thirst, and we once halted beside a fillet of clear spring water to have a draught. It seemed powerless to quench the drought which beset me. We reached a châlet; milking time was at hand, at our request a smart young Senner caught up a pail, and soon returned with it full of delicious

Ascent of the Weisshorn

milk. It was poured into a small tub. With my two hands I seized the two ends of a diameter of this vessel, gave it the necessary inclination, and stooping down, with a concentration of purpose which I had rarely before exerted, I drew the milk into me. Thrice I returned to the attack before that insatiate thirst gave way. The effect was astonishing. The liquid appeared to lubricate every atom of my body, and its fragrance to permeate my brain. I felt a growth of strength at once commence within me; all anxiety as to physical power with reference to the work in hand soon vanished, and before retiring to rest I was able to say to Benen, 'Go where thou wilt to-morrow, and I will follow thee.'

Two hours' additional climbing brought us to our bivouac. A ledge of rock jutted from the mountain side, and formed an overhanging roof. On removing the stones from beneath it, a space of comparatively dry clay was laid bare. This was to be my bed, and to soften it Wenger considerately stirred it up with his axe. The position was excellent, for lying upon my left side I commanded the whole range of Monte Rosa, from the Mischabel to the Breithorn. We were on the edge of an amphitheatre. Beyond the Schallenbach was the stately Mettelhorn. A row of eminent peaks swept round to the right, linked by lofty ridges of cliffs, thus forming the circus in which the Schallenberg Glacier originated. They were, however, only a spur cast out from the vaster Weisshorn, the cone of which was not visible from our dormitory. I wished to examine it, and in company with Benen skirted the mountain for half-an-hour, until the whole colossal pyramid stood facing us. When I first looked at it my hope sank, but both of us gathered confidence from a more lengthened gaze. The mountain is a pyramid with three faces, the intersections of which form three sharp edges or *arêtes*. The end of the eastern *arête* was nearest to us, and on it our attention was principally fixed. A couloir led up to it filled with snow, which Benen, after having examined it with the telescope, pronounced 'furchtbar steil.' This slope was cut across by a bergschrund, which we also carefully examined, and finally, Benen decided on the route to be pursued next morning. A chastened hope was predominant in both our breasts as we returned to our shelter.

Water was our first necessity; it seemed everywhere, but there was none to drink. It was locked to solidity in the ice and snow. The sound of it came booming up from the Vispbach, as it broke into foam or rolled its boulders over its waterworn bed; and the swish of many a minor streamlet mingled with the muffled roar of the large one. Benen set out in search of the precious liquid, and after a long absence returned with a jug and panful. I had been particular in including tea in our list of provisions; but on opening the parcel we found it half green, and not to be indulged in at a moment when the main object of one's life was to get an hour's sleep. We rejected the tea and made coffee instead. At our evening meal the idea of toasting our cheese occurred to Wenger, who is a man rich in expedients of all kinds. He turned the section of a large cheese towards the flame of our pine fire; it fizzed and blistered and turned viscous, and the toasted surface being removed was consumed with relish by us all. Our meal being ended and our beds arranged, by the help of Benen, I introduced myself into my two sacks in succession, and placed a knapsack beneath my head for a pillow. The talk now ceased and sleep became the object of our devotions.

But the goddess flies most shyly where she is most intensely wooed, still I think she touched my eyes gently once or twice during the night. The sunset had been unspeakably grand, steeping the zenith in violet, and flooding the base of the heavens with crimson light. Immediately opposite to us, on the other side of the valley of St. Nicholas, rose the Mischabel, with its two great peaks, the Grubenhorn and the Täschhorn, each barely under 15,000 feet in height. Next came the Alphubel, with its flattened crown of snow; then the Alleleinhorn and Rympfischhorn encased in glittering enamel; then the Cima di Jazzi; next the mass of Monte Rosa, with nothing competent to cast a shadow between it and the sun, and

Ascent of the Weisshorn

consequently flooded with light from bottom to top. The face of the Lyskamm turned towards us was for the most part shaded, but here and there its projecting portions jutted forth like red-hot embers as the light fell upon them. The 'Twins' were most singularly illuminated; across the waist of each of them was drawn a black bar produced by the shadow of a corner of the Breithorn, while their white bases and whiter crowns were exposed to the sunlight. Over the rugged face of the Breithorn itself the light fell as if in splashes, igniting its glaciers and swathing its black crags in a layer of transparent red. The Mettelhorn was cold, so was the entire range over which the Weisshorn ruled as king, while the glaciers which they embraced lay grey and ghastly in the twilight shade.

The sun is going, but not yet gone; while up the arch of the opposite heavens, the moon, within one day of being full, is hastening to our aid. She finally appears exactly behind the peak of the Rympfischhorn: the cone of the mountain being projected for a time as a triangle on the disc. Only for a moment, however; for the queenly orb sails aloft, clears the mountain, and bears splendidly away through the tinted sky. The motion was quite visible, and resembled that of a vast balloon. As the day approached its end the scene assumed the most sublime aspect. All the lower portions of the mountains were deeply shaded, while the loftiest peaks, ranged upon a semicircle, were fully exposed to the sinking sun. They seemed pyramids of solid fire, while here and there long stretches of crimson light drawn over the higher snowfields linked the glorified summits together. An intensely illuminated geranium flower seems to swim in its own colour which apparently surrounds the petals like a layer, and defeats by its lustre any attempt of the eye to seize upon the sharp outline of the leaves. A similar effect was here observed upon the mountains; the glory did not seem to come from them alone, but seemed also effluent from the air around them. This gave them a certain buoyancy which suggested entire detachment from the earth. They swam in splendour, which intoxicated the soul, and I will not now repeat in my moments of soberness the extravagant analogies which then ran through my brain. As the evening advanced, the eastern heavens low down assumed a deep purple hue, above which, and blending with it by infinitesimal gradations, was a belt of red, and over this again zones of orange and violet. I walked round the corner of the mountain at sunset, and found the western sky glowing with a more transparent crimson than that which overspread the east. The crown of the Weisshorn was embedded in this magnificent light. After sunset the purple of the east changed to a deep neutral tint, and against the faded red which spread above it, the sun-forsaken mountains laid their cold and ghastly heads. The ruddy colour vanished more and more; the stars strengthened in lustre, until finally the moon and they held undisputed possession of the blue grey sky.

I lay with my face turned towards the moon until it became so chilled that I was forced to protect it by a light handkerchief. The power of blinding the eyes is ascribed to the moonbeams, but the real mischief is that produced by radiation from the eyes into clear space, and the inflammation consequent upon the chill. As the cold increased I was fain to squeeze myself more and more underneath my ledge, so as to lessen the space of sky against which my body could radiate. Nothing could be more solemn than the night. Up from the valley came the low thunder of the Vispbach. Over the Dom flashed in succession the stars of Orion, until finally the entire constellation hung aloft. Higher up in heaven was the moon, and her rays as they fell upon the snow-fields and pyramids were sent back in silvery lustre by some, while others remained dull. These, as the orb sailed round, came duly in for their share of the glory. The Twins caught it at length and retained it long, shining with a pure spiritual radiance while the moon continued to ride above the hills.

I looked at my watch at 12 o'clock; and a second time at 2 A.M. The moon was then just touching the crest of the Schallenberg, and we were threatened with

Ascent of the Weisshorn 233

the withdrawal of her light. This soon occurred. We rose at 2¼ A.M., consumed our coffee, and had to wait idly for the dawn. A faint illumination at length overspread the west, and with this promise of the coming day we quitted our bivouac at 3½ A.M. No cloud was to be seen; as far as the weather was concerned we were sure to have fair play. We rounded the shingly shoulder of the mountain to the edge of a snow-field, but before entering upon it I disburthened myself of my strong shooting jacket, and left it on the mountain side. The sunbeams and my own exertion would, I knew, keep me only too warm during the day. We crossed the snow, cut our way through a piece of entangled glacier, reached the bergschrund, and passed it without a rope. We ascended the frozen snow of the couloir by steps, but soon diverged from it to the rocks at our right, and scaled them to the end of the eastern *arête* of the mountain.

Here a saddle of snow separates us from the next higher rocks. With our staff-spikes at one side of the saddle, we pass by steps cut upon the other. The snow is firmly congealed. We reach the rocks, which we find hewn into fantastic turrets and obelisks, while the loose chips of this colossal sculpture are strewn confusedly upon the ridge. Amid the chips we cautiously pick our way, winding round the towers or scaling them amain. From the very first the work is heavy, the bending, twisting, reaching, and drawing up, calling upon all the muscles of the frame. After two hours of this work we halt, and looking back we observe two moving objects on the glacier below us. At first we take them to be chamois, but they are instantly pronounced men, and the telescope at once confirms this. The leader carries an axe, and his companion a knapsack and alpenstock. They are following our traces, losing them apparently now and then, and waiting to recover them. Our expedition had put Randa in a state of excitement, and some of its best climbers had come to Benen and urged him to take them with him. But this he did not deem necessary, and now here were two of them determined to try the thing on their own account; and perhaps to dispute with us the honour of the enterprise. On this point, however, our uneasiness was small.

Resuming our gymnastics, the rocky staircase led us to the flat summit of a tower, where we found ourselves cut off from a similar tower by a deep gap bitten into the mountain. Retreat appeared inevitable, but it is wonderful how many ways out of difficulty open to a man who diligently seeks them. The rope is here our refuge. Benen coils it round his waist, scrapes along the surface of the rock, fixes himself on a ledge, where he can lend me a helping hand. I follow him, Wenger follows me, and in a few minutes all three of us stand in the middle of the gap. By a kind of screw motion we twist ourselves round the opposite tower, and reach the *arête* behind it. Work of this kind, however, is not to be performed by the day, and with a view of sparing our strength, we quit the *arête* and endeavour to get along the southern slope of the pyramid. The mountain is here scarred by longitudinal depressions which stretch a long way down it. These are now filled with clear hard ice, produced by the melting and refreezing of the snow. The cutting of steps across these couloirs proves to be so tedious and fatiguing, that I urge Benen to abandon it and try the *arête* once more. By a stout tug we regain the ridge and work along it as before. Here and there from the northern side the snow has folded itself over the crags, and along it we sometimes work upward. The *arête* for a time has become gradually narrower, and the precipices on each side more sheer. We reach the end of one of the subdivisions of the ridge, and find ourselves separated from the next rocks by a gap about twenty yards across. The *arête* here has narrowed to a mere wall, which, however, as rock would present no serious difficulty. But upon the wall of rock is placed a second wall of snow, which dwindles to a knife edge at the top. It is white and pure, of very fine grain, and a little moist. How to pass this snow catenary I knew not, for I had no idea of a human foot trusting itself upon so frail a support. Benen's practical sagacity

Ascent of the Weisshorn

was, however, greater than mine. He tried the snow by squeezing it with his foot, and to my astonishment commenced to cross. Even after the pressure of his feet the space he had to stand on did not exceed a handbreadth. I followed him, exactly as a boy walking along a horizontal pole, with toes turned outwards. Right and left the precipices were appalling; but the sense of power on such occasions is exceedingly sweet. We reached the opposite rock, and here a smile rippled over Benen's countenance as he turned towards me. He knew that he had done a daring thing, though not a presumptuous one. 'Had the snow,' he said, 'been less perfect, I should not have thought of attempting it, but I knew after I had set my foot upon the ridge that we might pass without fear.'

It is quite surprising what a number of things the simple observation made by Faraday, in 1846, enables us to explain. Benen's instinctive act is justified by theory. The snow was fine in grain, pure and moist. When pressed, the attachments of its granules were innumerable, and their perfect cleanness enabled them to freeze together with a maximum energy. It was this freezing together of the particles at innumerable points which gave the mass its sustaining power. Take two fragments of ordinary table ice and bring them carefully together, you will find that they freeze and cement themselves at their place of junction; or if two pieces float in water, you can bring them together, when they instantly freeze, and by laying hold of either of them gently, you can drag the other after it through the water. Imagine such points of attachment distributed without number through a mass of snow. The substance becomes thereby a semi-solid instead of a mass of powder. My guide, however, unaided by any theory, did a thing from which I, though backed by all the theories in the world, should have shrunk in dismay.

After this we found the rocks on the ridge so shaken to pieces that it required the greatest caution to avoid bringing them down upon us. With all our care, however, we sometimes dislodged vast masses which leaped upon the slope adjacent, loosened others by their shock, these again others, until finally a whole flight of them would escape, setting the mountain in a roar as they whizzed and thundered along its side to the snow-fields 4000 feet below us. The day is hot, the work hard, and our bodies are drained of their liquids as by a Turkish bath. The perspiration trickles down our faces, and drops profusely from the projecting points. To make good our loss we halt at intervals where the melted snow forms a liquid vein, and quench our thirst. We possess, moreover, a bottle of champagne, which, poured sparingly into our goblets on a little snow, furnishes Wenger and myself with many a refreshing draught. Benen fears his eyes, and will not touch champagne. The less, however, we rest the better, for after every pause I find a certain unwillingness to renew the toil. The muscles have become set, and some minutes are necessary to render them again elastic. But the discipline is first-rate for both mind and body. There is scarcely a position possible to a human being which, at one time or another during the day, I was not forced to assume. The fingers, wrist, and forearm, were my main reliance, and as a mechanical instrument the human hand appeared to me this day in a light which it never assumed before. It is a miracle of constructive art.

We were often during the day the victims of illusions regarding the distance which we had to climb. For the most part the summit was hidden from us, but on reaching the eminences it came frequently into view. After three hours spent on the *arête*, about five hours that is, subsequent to starting, the summit was clearly in view; we looked at it over a minor summit, which gave it an illusive proximity. 'You have now good hopes,' I remarked, turning to Benen. 'Not only good hopes,' he replied, 'but I do not allow myself to entertain the idea of failure.' Well, six hours passed on the *arête*, each of which put in its inexorable claim to the due amount of mechanical work; the lowering and the raising of three human bodies through definite spaces, and at the end of this time we found ourselves apparently no nearer to

Ascent of the Weisshorn

the summit than when Benen's hopes cropped out in confidence. I looked anxiously at my guide as he fixed his weary eyes upon the distant peak. There was no confidence in the expression of his countenance; still I do not believe that either of us entertained for a moment the thought of giving in. Wenger complained of his lungs, and Benen counselled him several times to stop and let him and me continue the ascent; but this the Oberland man refused to do. At the commencement of a day's work I often find myself anxious, if not timid; but this feeling vanishes when I become warm and interested. When the work is very hard we become callous and sometimes stupefied by the incessant knocking about. This was my case at present, and I kept watch lest my indifference should become carelessness. I supposed repeatedly a case where a sudden effort might be required of me, and felt all through that I had a fair residue of strength to fall back upon. I tested this conclusion sometimes by a spurt; flinging myself suddenly from rock to rock, and thus proved my condition by experiment instead of relying on opinion. An eminence in the ridge which cut off the view of the summit was now the object of our exertions. We reached it; but how hopelessly distant did the summit appear! Benen laid his face upon his axe for a moment; a kind of sickly despair was in his eye as he turned to me, remarking, 'Lieber Herr, die Spitze ist noch sehr weit oben.'

Lest the desire to gratify me should urge him beyond the bounds of prudence, I said to Benen that he must not persist on my account, if he ceased to feel confidence in his own powers; that I should cheerfully return with him the moment he thought it no longer safe to proceed. He replied that though weary he felt quite sure of himself, and asked for some food. He had it, and a gulp of wine, which mightily refreshed him. Looking at the mountain with a firmer eye, he exclaimed, 'Herr! wir müssen ihn haben,' and his voice, as he spoke, rung like steel within my heart. I thought of Englishmen in battle, of the qualities which had made them famous, it was mainly the quality of not knowing when to yield; of fighting for duty even after they had ceased to be animated by hope. Such thoughts had a dynamic value, and helped to lift me over the rocks. Another eminence now fronted us, behind which, how far we knew not, the summit lay. We scaled this height, and above us, but clearly within reach, a silvery pyramid projected itself against the blue sky. I was assured ten times by my companions that it was the highest point before I ventured to stake my faith upon the assertion. I feared that it also might take rank with the illusions which had so often beset our ascent, and shrunk from the consequent moral shock. Towards the point, however, we steadily worked. A large prism of granite, or granitic gneiss, terminated the *arête*, and from it a knife edge of pure white snow ran up to a little point. We passed along the edge, reached that point, and instantly swept with our eyes the whole range of the horizon. The crown of the Weisshorn was underneath our feet.

The long pent feelings of my two guides found vent in a wild and reiterated cheer. Benen shook his arms in the air and shouted as a Valaisian, while Wenger chimed in with the shriller yell of the Oberland. We looked along the *arête*, and far below perched on one of its crags, could discern the two Randa men. Again and again the roar of triumph was sent down to them. They had accomplished but a small portion of the ridge, and soon after our success they wended their way homewards. They came, willing enough, no doubt, to publish our failure had we failed; but we found out afterwards that they had been equally strenuous in announcing our success; they had seen us, they affirmed, like three flies upon the summit of the mountain. Both men had to endure a little persecution for the truth's sake, for nobody in Randa would believe that the Weisshorn could be scaled, and least of all by a man who for two days previously had been the object of Philomène, the waiter's, constant pity, on account of the incompetence of his stomach to accept all that she offered for its acceptance. The energy of conviction with which the men gave their evidence had, however, convinced the most sceptical before we arrived ourselves.

Ascent of the Weisshorn

Benen wished to leave some outward and visible sign of our success on the summit. He deplored having no flag; but as a substitute it was proposed that he should knock the head off his axe, use the handle as a flagstaff, and surmount it by a red pocket-handkerchief. This was done, and for some time subsequently the extempore banner was seen flapping in the wind. To his extreme delight, it was shown to Benen himself three days afterwards by my friend Mr. Galton from the Riffel Hotel. But you will desire to know what we saw from the summit, and this desire I am sorry to confess my total incompetence to gratify. I remember the picture, but cannot analyse its parts. Every Swiss tourist is acquainted with the Weisshorn. I have long regarded it as the noblest of all the Alps, and many, if not most other travellers, have shared this opinion. The impression it produces is in some measure due to the comparative isolation with which its cone juts into the heavens. It is not masked by other mountains, and all around the Alps its final pyramid is in view. Conversely the Weisshorn commands a vast range of prospect. Neither Benen nor myself had ever seen anything at all equal to it. The day, moreover, was perfect; not a cloud was to be seen; and the gauzy haze of the distant air, though sufficient to soften the outlines and enhance the colouring of the mountains, was far too thin to obscure them. Over the peaks and through the valleys the sunbeams poured, unimpeded save by the mountains themselves, which in some cases drew their shadows in straight bars of darkness through the illuminated air. I had never before witnessed a scene which affected me like this. Benen once volunteered some information regarding its details, but I was unable to hear him. An influence seemed to proceed from it direct to the soul; the delight and exultation experienced were not those of Reason or of Knowledge, but of BEING:—I was part of it and it of me, and in the transcendent glory of Nature I entirely forgot myself as man. Suppose the sea waves exalted to nearly a thousand times their normal height, crest them with foam, and fancy yourself upon the most commanding crest, with the sunlight from a deep blue heaven illuminating such a scene, and you will have some idea of the form under which the Alps present themselves from the summit of the Weisshorn. East, west, north, and south, rose those 'billows of a granite sea,' back to the distant heaven, which they hacked into an indented shore. I opened my note-book to make a few observations, but I soon relinquished the attempt. There was something incongruous, if not profane, in allowing the scientific faculty to interfere where silent worship was the 'reasonable service.'

CHAPTER VII

THE DESCENT

' He clasps the crag with hooked hands.'

WE had been ten hours climbing from our bivouac to the summit, and it was now necessary that we should clear the mountain before the close of day. Our muscles were loose and numbed, and unless extremely urged declined all energetic tension: the thought of our success, however, ran like a kind of wine through our fibres and helped us down. We once fancied that the descent would be rapid, but it was far from it. Benen, as in ascending, took the lead: he slowly cleared each crag, paused till I joined him, I pausing till Wenger joined me, and thus one or other of us was always in motion. Benen shows a preference for the snow where he can choose it, while I hold on to the rocks where my hands can assist my feet. Our muscles are sorely tried by the twisting round the splintered turrets of the *arête*, and we resolve to escape from it when we can; but a long, long stretch of the ridge must be passed before we dare to swerve from it. We are roused from our stupefaction at times by the roar of the stones which we have loosed from the ridge, and sent leaping down the mountain. The snow catenary is attained, and we recross it. Soon afterwards we quit the ridge and try to get obliquely along the slope of the mountain. The face of the pyramid is here scarred by couloirs, of which the deeper and narrower ones are filled with ice, while the others are highways to the bottom of the mountain for the rocks quarried by the weather above. Steps must be cut in the ice, but the swing of the axe is very different now from what it was in the morning. Still, though Benen's blows descend with the deliberateness of a man whose fire is half-quenched, they fall with sufficient power, and the needful cavities are soon formed. We retrace our morning steps over some of the slopes. No word of warning was uttered here as we ascended, but now Benen's admonitions were frequent and emphatic,—'Take care not to slip.' I looked down the slopes; they seemed fearfully long, and those whose ends we could see were continued by rocks over which it would be the reverse of comfortable to be precipitated. I imagined, however, that even if a man slipped he would be able to arrest his descent; but Benen's response when I stated this opinion was very prompt,—'No! it would be utterly impossible. If it were snow you might do it, but it is pure ice, and if you fall you will lose your senses before you can use your axe.' I suppose he was right. At length we turn directly downwards, and work along one of the ridges which are here drawn parallel to the line of steepest fall. We first drop cautiously from ledge to ledge. At one place Benen clings for a considerable time to a face of rock, casting out feelers of leg and arm, and desiring me to stand still. I do not understand the difficulty, for the rock though steep is by no means vertical. I fasten myself to it, but Benen is now on a ledge below, waiting to receive me. The spot on which he stands is a little rounded protuberance sufficient to afford him footing, but over which the slightest momentum would have carried him. He knew this, and hence his caution in descending. Soon after this we quit our ridge and drop into a couloir to the left of it. It is dark and damp with trickling water. The rope hampers us, and I propose its abandonment. We disencumber ourselves, and find our speed greatly increased. In some places the rocks are worn to a powder, along which we shoot by glissades. We swerve again to the left; cross a ridge, and get into another and drier couloir. The last one was dangerous, as the water exerted a constant sapping action upon the rocks. From our new position we could hear the clatter of stones descending the gully which we had just forsaken. Wenger, who had brought up the rear during the day, is now sent to the front; he has not Benen's power, but his legs are long and his descent rapid. He scents out the way,

The Descent

which becomes more and more difficult. He pauses, observes, dodges, but finally comes to a dead stop on the summit of a precipice, which sweeps like a rampart round the mountain. We move to the left, and after a long detour succeed in rounding the rocky wall. Again straight downwards. Half-an-hour brings us to the brow of a second precipice, which is scooped out along its centre so as to cause the brow to overhang. I see chagrin in Benen's face: he turns his eyes upwards, and I fear mortally that he is about to propose a re-ascent to the *arête*. He had actually thought of doing so, but it was very questionable whether our muscles could have responded to such a demand. While we stood pondering here, a deep and confused roar attracted our attention. From a point near the summit of the Weisshorn, a rock had been discharged; it plunged down a dry couloir, raising a cloud of dust at each bump against the mountain. A hundred similar ones were immediately in motion, while the spaces between the larger masses were filled by an innumerable flight of smaller stones. Each of them shakes its quantum of dust in the air, until finally the avalanche is enveloped in a vast cloud. The clatter of this devil's cavalry was stunning. Black masses of rock emerged here and there from the cloud, and sped through the air like flying fiends. Their motion was not one of translation merely, but they whizzed and vibrated in their flight as if urged by wings. The clang of echoes resounded from side to side, from the Schallenberg to the Weisshorn and back, until finally the whole troop came to rest, after many a deep-sounding thud in the snow, at the bottom of the mountain. This stone avalanche was one of the most extraordinary things I had ever witnessed, and in connection with it, I would draw the attention of future climbers to the danger which would infallibly beset any attempt to ascend the Weisshorn from this side, except by one of its *arêtes*. At any moment the mountain side may be raked by a fire as deadly as that of cannon.

After due deliberation we move along the precipice westward, I fearing that each step forward is but plunging us into deeper difficulty. At one place, however, the precipice bevels off to a steep incline of smooth rock. Along this runs a crack, wide enough to admit the fingers, and sloping obliquely down to the lower glacier. Each in succession grips the rock and shifts his body sideways parallel to the fissure, until he comes near enough to the glacier to let go and slide down by a rough glissade. We afterwards pass swiftly along the glacier, sometimes running, and, on the steeper slopes, by sliding, until we are pulled up for the third time by a precipice which seems actually worse than either of the others. It is quite sheer, and as far as I can see right or left altogether hopeless. I fully expected to hear Benen sound a retreat, but to my surprise both men turned without hesitation to the right, which took us away from our side of the mountain. I felt desperately blank, but I could notice no expression of dismay in the countenance of either of the men. They observed the moraine matter over which we walked, and at length one of them exclaimed, 'Da sind die Spuren,' lengthening his strides at the same moment. We look over the brink at intervals, and at length discover what appears to be a mere streak of clay on the face of the precipice. We get round a corner, and find footing on this streak. It is by no means easy, but to hard-pushed men it is a deliverance. The streak vanishes, and we must scrape down the rock. This fortunately is rough, so that by pressing the hands against its rounded protuberances, and sticking the boot-nails against its projecting crystals, we let our bodies gradually down. We thus reach the bottom; a deep cleft separates the glacier from the precipice, this is crossed, and we are now free men, clearly placed beyond the last bastion of the mountain.

I could not repress an expression of admiration at the behaviour of my men. The day previous to my arrival at Randa they had been up to examine the mountain, when they observed a solitary chamois moving along the base of this very precipice, and making several ineffectual attempts to get up it. At one place the creature succeeded; this spot they marked in their

The Descent

memories as well as they could, and when they reached the top of the precipice they sought for the traces of the chamois, found them, and were guided by them to the only place where escape in any reasonable time was possible. Our way is now clear; over the glacier we cheerfully march, and pass from the ice just as the moon and the eastern sky contribute about equally to the illumination. Wenger makes direct for our resting-place and packs up our things, while Benen and myself try to descend towards the châlet. Clouds gather round the Rympfischhorn and intercept the light of the moon. We are often at a loss, and wander half-bewildered over the Alp. At length the welcome tinkle of cowbells is heard in the distance, and guided by them we reach the châlet a little after 9 P.M. The cows had been milked and the milk disposed of, but the men managed to get us a moderate draught. Thus refreshed we continue the descent, and are soon amid the pines which clothe the mountain facing Randa. A light glimmers from the window of the hotel; we conclude that they are waiting for us; it disappears, and we infer that they have gone to bed. Wenger is sent on to order some food; I was half-famished, for my nutriment during the day consisted solely of a box of meat lozenges given to me by Mr. Hawkins. Benen and myself descend the mountain deliberately, and after many windings emerge upon the valley, cross it, and reach the hotel a little before 11 P.M. I had a basin of broth, *not* made according to Liebig, and a piece of mutton boiled probably for the seventh time. Fortified by these, and comforted by a warm footbath, I went to bed, where six hours' sound sleep chased away every memory of the Weisshorn save the pleasant ones. I was astonished to find the loose atoms of my body knitted so firmly together by so brief a rest. Up to my attempt upon the Weisshorn I had felt more or less dilapidated, but here all weakness ended. My fibres assumed more and more the tenacity of steel, and during my subsequent stay in Switzerland I was unacquainted with infirmity. If you, my friend, should ask me why I incur such labour and such risk, here is *one* reply.

The height of the Weisshorn is fourteen thousand eight hundred and thirteen feet. Height, however, is but one element in the difficulty of a mountain. Monte Rosa, for example, is higher than the Weisshorn, but the difficulty of the former is small in comparison to that of the latter.

CHAPTER VIII

THE MOTION OF GLACIERS

' The god that made New Hampshire,
Taunted the lofty land
With little men.'

IT is impossible for a man with his eyes open to climb a mountain like the Weisshorn without having his knowledge augmented in many ways. The mutations of the atmosphere, the blue zenith and the glowing horizon, rocks, snow, and ice; the wondrous mountain world into which he looks, and which refuses to be encompassed by a narrow brain:—these are objects at once poetic and scientific, and of such plasticity that every human soul can fashion them according to its own needs. It is not my object to dwell on these things at present, but I made one little observation in descending the Weisshorn to which I should like, in a more or less roundabout way, to direct your attention.

The wintry clouds, as you know, drop spangles on the mountains. If the thing occurred once in a century, historians would chronicle and poets would sing of the event; but Nature, prodigal of beauty, rains down her hexagonal ice-stars year by year, forming layers yards in thickness. The summer sun thaws and partially consolidates the mass. Each winter's fall is covered by that of the ensuing one, and thus the snow layer of every year has to sustain an annually augmented weight. It is more and more compacted by the pressure, and ends by being converted into the ice of a true glacier, which stretches its frozen tongue far down below the limits of perpetual snow.

The glaciers move, and through valleys they move like rivers. 'Between the Mer de Glace and a river,' writes Rendu, 'there is a resemblance so complete that it is impossible to find in the latter a circumstance which does not exist in the former.' A cork when cast upon a stream, near its centre, will move more quickly than when thrown near the sides, for the progress of the stream is retarded by its banks. And as you and your guide stood together on the solid waves of that Amazon of ice you were borne resistlessly along. You saw the boulders perched upon their frozen pedestals; these were the spoils of distant hills, quarried from summits far away, and floated to lower levels like timber logs upon the Rhone. As you advanced towards the centre you were carried down the valley with an ever-augmenting velocity. You felt it not—he felt it not—still you were borne down with a velocity which, if continued, would amount to 1000 feet a year.

And could you have cast a log into the solid mass and determined the velocity of its deeper portions, you would have learned that the ice-river, like the liquid one, is retarded by its bed; that the surface of the glacier moves more quickly than the bottom. You remember also the shape of that other glacier, where you passed along an ice crest six inches wide, with chasms of unsounded depth right and left. You never trembled; but you once swayed, and the guide bruised your arm by the pressure of his fingers. I told you when informed of this, that the shape of the valley was to blame. My meaning was this: the valley formed a curve at the place, and you stood upon the convex side of the glacier. This side was moving more speedily than the opposite one, thereby tearing itself more fiercely asunder. Hence arose the chasms which you then encountered. At this place the eastern side of the glacier moved more quickly than the western one. Higher up, the valley bent in the opposite direction, and there the western side moved quickest. Thus, exactly as in the Ribble and the Aire, and the Wye and the Thames; the place of swiftest motion of the glacier shifted from side to side in obedience to the curvature of the valley.

To a Savoyard priest, who, I am happy to say, afterwards became a bishop, we are indebted for the first

The Motion of Glaciers

clear enunciation of the truth that a glacier moves as a river; an idea which, as you know, was subsequently maintained with energy and success by a distinguished countryman of our own. Rendu called the portion of the glacier with which you are acquainted, and which is confined between banks of mountains, 'the *flowing* glacier' (*Glacier d'écoulement*), associating with the term 'flowing,' the definite physical idea which belongs to it; and he called the basin, or the plateau, in which the snows which fed the lower ice-stream were collected the '*reservoir.*' He assigned a true origin to the glacier, a true progress, and a true end; and yet you, acquainted as you are with Alpine literature, and warmly as you were interested in the discussions to which that literature has given birth; you, I say, had actually forgotten the existence of this bishop, and required time to persuade yourself of his merits, when his claims were introduced in your presence before a society of friends three years ago.

Some have blamed me, and some have praised me, for the part which I have acted towards this man's memory. In one distinguished, but not disinterested quarter, I have been charged with prejudice and littleness of spirit; to which charge I have nothing to reply. A peaceable man when thus assailed will offer no resistance. But you, my friend, know how light a value I set on my scientific labours in the Alps. Indeed, I need them not. The glaciers and the mountains have an interest for me beyond their scientific ones. They have been to me well-springs of life and joy. They have given me royal pictures and memories which can never fade. They have made me feel in all my fibres the blessedness of perfect manhood, causing mind, and soul, and body, to work together with a harmony and strength unqualified by infirmity or ennui. They have raised my enjoyments to a higher level, and made my heart competent to cope even with yours in its love of Nature. This has been the bounty of the Alps to me. And it is sufficient. I should look less cheerily into the future did I not hope to micrify, by nobler work, my episode upon the glaciers. On it I shall never found the slightest claim of my own; but I do claim the right, and shall ever exercise it, of doing my duty towards my neighbour, and of giving to forgotten merit its award. I have done no more. Let it be made clear that I have wronged any man by false accusation, and Zacchæus was not more prompt than I shall be to make restitution. We may have all erred more or less in connection with this question; but had a little more chivalry been imported into its treatment twenty years ago, these personal discussions would not now associate themselves with the glaciers of the Alps.

But the glaciers have a motion besides that which they owe to the *quasi* plasticity of their own masses. Ice is slippy; ice is fusible; and in dead winter water flows along the glacier's bed. In dead winter the under surface of the glacier is wearing away. The glacier slides bodily over its rocky bed. 'Prove this,' you have a right to retort. Well, here is one proof. You have heard me speak of the fluted rocks of the Grimsel; you have heard of the ancient glaciers of England, Scotland, Ireland, and Wales. Killarney, to which I have already referred, affords magnificent examples of ancient glacier action. No man with the slightest knowledge of the glacier operations of to-day could resist the conclusion, that the Black Valley of Killarney was once filled by a glacier fed by the snows from Magillicuddy's Reeks; that the 'Cannon Rock,' the 'Man of War,' the 'Giant's Coffin,' and other masses fantastically named, were moulded to their present shapes by a glacier which once held possession of the hollow now filled by the 'Upper Lake.' No man can resist the evidence of glacier action on the flanks of Snowdon, and round about the slopes of Scawfell Pike and Great Gable. And what is the nature of the evidence which thus refuses to be gainsaid? Simply the scoring and polishing and fluting of the rocks to which I have so often referred. Although executed ages ago, they are as fresh and unmistakable as if they had been executed last year; and to leave such marks and tokens behind it, the glacier must have slidden over its bed.

The Motion of Glaciers 251

Here, then, is one proof of glacier sliding which was urged many years ago, and I think it is satisfactory. But not only does the glacier act upon the rocks, but the rocks must of necessity act upon the under surface of the glacier; and could we inspect this, we should assuredly find proof of sliding. This proof exists, and I am unable to state it in clearer language than that employed in the following letter which I have already published.

'Many years ago Mr. William Hopkins of Cambridge, pointed to the state of the rocks over which glaciers had passed as conclusive evidence that these vast masses of ice move bodily along their beds. Those rocks are known to have their angles rasped off, and to be fluted and scarred by the ice which has passed over them. Such appearances, indeed, constitute the entire evidence of the former existence of glaciers in this and other countries, discussed in the writings of Venetz, Charpentier, Agassiz, Buckland, Darwin, Ramsay, and other eminent men.

'I have now to offer a proof of the sliding of the ice exactly complementary to the above. Suppose a glacier to be a plastic mass, which did not slide, and suppose such a glacier to be turned upside down, so as to expose its under surface; that surface would bear the impression of its bed, exactly as melted wax bears the impression of a seal. The protuberant rocks would make hollows of their own shape in the ice, and the depressions of the bed would be matched by protuberances of their own shape on the under surface of the glacier. But, suppose the mass to slide over its bed, these exact impressions would no longer exist; the protuberances of the bed would then form longitudinal furrows, while the depressions of the bed would produce longitudinal ridges. From the former state of things we might infer that the bottom of the glacier is stationary, while from the latter we should certainly infer that the whole mass slides over its bed.

'In descending from the summit of the Weisshorn on the 19th of August last I found, near the flanks of one of its glaciers, a portion of the ice completely roofing a hollow, over which it had been urged without being squeezed into it. A considerable area of the under surface of the glacier was thus exposed, and the ice of that surface was more finely fluted than ever I have observed rocks to be. Had the tool of a cabinetmaker passed over it, nothing more regular and beautiful could have been executed. Furrows and ridges ran side by side in the direction of the motion, and the deeper and larger ones were chased by finer lines, produced by the smaller and sharper asperities of the bed. The ice was perfectly unweathered, and the white dust of the rocks over which it had passed, and which it had abraded in its passage, still clung to it. The fact of sliding has been hitherto inferred from the action of the glacier upon the rocks; the above observation leads to the same inference from the action of the rocks upon the glacier. As stated at the outset, it is the complementary proof that the glacier moves bodily over its bed.'

CHAPTER IX

SUNRISE ON THE PINES

*The sunbeam gave me to the sight
The tree adorned the formless light.'*

I MUST here mention a beautiful effect which I observed from Randa on the morning of the 18th of August. The valley of St. Nicholas runs nearly north and south, and the ridge which flanks it to the east is partially covered with pines; the trees on the summit of this ridge as you look at them from the valley being projected against the sky. What I saw was this: as the sun was about to rise I could trace upon the meadows in the valley the outline of the shadow of the ridge which concealed him, and I could walk along the valley so as to keep myself quite within the shadow of the mountain. Suppose me just immersed in the shadow: as I moved along, successive pine-trees on the top of the ridge were projected on that portion of the heavens where the sun was about to appear, and every one of them assumed in this position a perfect silvery brightness. It was most interesting to observe, as I walked up or down the valley, tree after tree losing its opacity and suddenly robing itself in glory. Benen was at mass at the time, and I drew Wenger's attention to the effect. He had never observed it before. I never met a guide who had—a fact to be explained by the natural repugnance of the eyes to be turned towards a sky of dazzling brightness. Professor Necker was the first who described this effect, and I have copied his description in 'the Glaciers of the Alps.' The only difference between his observation and mine is, that whereas he saw the *stems* of the trees also silver bright, I saw them drawn in dark streaks through the lustrous branches. The cause of the phenomenon I take to be this: You have often noticed the bright illumination of the atmosphere immediately surrounding the sun; and how speedily the brightness diminishes as your eye departs from the sun's edge. This brightness is mainly caused by the sunlight falling on the aqueous particles in the air, aided by whatever dust may be suspended in the atmosphere. If instead of aqueous particles fine solid particles were strewn in the air, the intensity of the light reflected from them would be greater. Now the spiculæ of the pine, when the tree is projected against the heavens, close to the sun's rim, are exactly in this condition; they are flooded by a gush of the intensest light, and reflect it from their smooth surfaces to the spectator. Every needle of the pine is thus burnished, appearing almost as bright as if it were cut out of the body of the sun himself. Thus the leaves and more slender branches shine with exceeding glory, while the surfaces of the thicker stems which are turned from the sun escape the light, and are drawn as dark lines through the brightness. Their diameters, however, are diminished by the irradiation from each side of them. I have already spoken of the lustre of thistle-down, in my book upon the Alps, and two days after the observation at Randa, I saw from Zermatt innumerable fragments of the substance floating at sunset in the western heaven, not far from the base of the Matterhorn. They gleamed like fragments of the sun himself. The lustre of the trees, then, I assume to be due to the same cause as the brilliancy of the heavens close to the sun; the superior intensity of the former being due to the greater quantity of light reflected from the solid spiculæ.

CHAPTER X

INSPECTION OF THE MATTERHORN

' By million changes skilled to tell
What in the Eternal standeth well,
And what obedient Nature can,
Is this colossal talisman.'

ON the afternoon of the 20th we quitted Randa, with a threatening sky overhead. The considerate Philomène compelled us to take an umbrella, which we soon found useful. The flood-gates of heaven were unlocked, while defended by our cotton canopy, Benen and myself walked arm in arm to Zermatt. I instantly found myself in the midst of a circle of pleasant friends, some of whom had just returned from a successful attempt upon the Lyskamm. On the 22nd quite a crowd of travellers crossed the Theodule Pass; and knowing that every corner of the hotel at Breuil would be taken up, I halted a day so as to allow the people to disperse. Breuil, as you know, commands a view of the south side of the Matterhorn; and it was now an object with me to discover, if possible, upon the true peak of this formidable mountain, some ledge or cranny, where three men might spend a night. The mountain may be accessible or inaccessible, but one thing seems certain, that starting from Breuil, or even from the chalets above Breuil, the work of reaching the summit is too much for a single day. But could a shelter be found amid the wild battlements of the peak itself, which would enable one to attack the obelisk at day-dawn, the possibility of conquest was so far an open question as to tempt a trial. I therefore sent Benen on to reconnoitre, purposing myself to cross the Theodule alone on the following day.

On the afternoon of the 22nd, I walked up to the Riffel, sauntering slowly, leaning at times on the head of my axe, or sitting down upon the grassy knolls, as my mood prompted. I have spoken with due reverence of external nature, still the magnificence of this is not always a measure of the traveller's joy. The joy is a polar influence made up of two complementary parts, the outward object, and the inward harmony with that object. Thus, on the hackneyed track to the Riffel, it is possible to drink the deepest delight from the contemplation of the surrounding scene. It was dinner-hour at the hotel above—dinner-hour at the hotel below, and there seemed to be but a single traveller on the way between them. The Matterhorn was all bare, and my vision ranged with an indefinable longing from base to summit over its blackened crags. The air which filled the valleys of the Oberland, and swathed in mitigated density the highest peaks, was slightly aqueous, though transparent, the watery particles forming so many *points d'appui* from which the sunbeams were scattered through surrounding space. The whole medium glowed as if with the red light of a distant furnace, and through it the outline of the mountains grandly loomed. The glow augmented as the sun sank, reached its maximum, paused, and then ran speedily down to a cold and colourless twilight.

Next morning at nine o'clock, with some scraps of information from the guides to help me on my way, I quitted the Riffel to cross the Theodule. I was soon followed by the domestic of the hotel; a very strong fellow, kept by M. Seiler as a guide up Monte Rosa. Benen had requested him to see me to the edge of the glacier, and he now joined me with this intention. He knew my designs upon the Matterhorn, and strongly deprecated them. 'Why attempt what is impossible?' he urged. 'What you have already accomplished ought to satisfy you, without putting your life in such certain peril. Only think, Herr, what will avail your ascent of the Weisshorn if you are smashed upon the Mont Cervin. Mein Herr!' he added with condensed emphasis, 'thun Sie es nicht.' The whole conversation was in fact a homily, the strong point of which was the utter uselessness of success on the one mountain if it were to be followed by annihilation on the other. We reached the ridge above the glacier, where,

Inspection of the Matterhorn

handing him a trinkgeld, which I had to force on his acceptance, I bade him good-bye, assuring him that I would submit in all things to Benen's opinion. He had the highest idea of Benen's wisdom, and hence the assurance sent him home comforted.

I was soon upon the ice, once more alone, as I delight to be at times. You have sometimes blamed me for going alone, and the right to do so ought to be earned by long discipline. As a habit I deprecate it; but sparingly indulged in, it is a great luxury. There are no doubt moods when the mother is glad to get rid of her offspring, the wife of her husband, the lover of his mistress, and when it is not well to keep them together. And so, at rare intervals, it is good for the soul to feel the influence of that 'society where none intrudes.' When your work is clearly within your power, when long practice has enabled you to trust your own eye and judgment in unravelling crevasses, and your own axe and arm in subduing their more serious difficulties, it is an entirely new experience to be alone amid those sublime scenes. The peaks wear a more solemn aspect, the sun shines with a more effectual fire, the blue of heaven is more deep and awful, the air seems instinct with religion, and the hard heart of man is made as tender as a child's. In places where the danger is not too great, but where a certain amount of skill and energy are required, the feeling of self-reliance is inexpressibly sweet, and you contract a closer friendship with the universe in virtue of your more intimate contact with its parts. The glacier to-day filled the air with low murmurs, which the sound of the distant moulins raised to a kind of roar. The débris rustled on the moraines, the smaller rivulets babbled in their channels, as they ran to join their trunk, and the surface of the glacier creaked audibly as it yielded to the sun. It seemed to breathe and whisper like a living thing. To my left was Monte Rosa and her royal court, to my right the mystic pinnacle of the Matterhorn, which from a certain point here upon the glacier attains its maximum sharpness. It drew my eyes towards it with irresistible fascination as it shimmered in the blue, too preoccupied with heaven, to think even with contempt on the designs of a son of earth to reach its inviolate crest.

Well, I crossed the Görner Glacier quite as speedily as if I had been professionally led. Then up the undulating slope of the Theodule Glacier with a rocky ridge to my right, over which I was informed a rude track led to the pass of St. Theodule. I am not great at finding tracks, and I missed this one, ascending until it became evident to me that I had gone too far. Near its higher extremity the crest of the ridge is cut across by three curious chasms, and one of these I thought would be a likely gateway through the ridge. I climbed the steep buttress of the spur and was soon in the fissure. Huge masses of rock were jammed into it, the presence of which gave variety to the exertion. I ascended along the angles between them and the cliffs to the left of them; the work was very pleasant, calling forth strength, but not exciting fear. From the summit the rocks sloped gently down to the snow, and in a few minutes the presence of broken bottles on the moraine showed me that I had hit upon the track over the pass. Upwards of twenty unhappy bees staggered against me on the way; tempted by the sun, or wafted by the wind, they had quitted the flowery Alps to meet torpor and death in the ice world above. From the summit I went swiftly down to Breuil, where I was welcomed by the host, welcomed by the waiter; loud were the expressions of content at my arrival; and I was informed that Benen had started early in the morning to 'promenade himself' around the Matterhorn.

I lay long upon the Alp, scanning crag and snow in search of my guide, and not doubting that his report would be favourable. You are already acquainted with the admirable account of our attempt on the Matterhorn drawn up by Mr. Hawkins, and from it you may infer that the ascent of this mountain is not likely to be a matter of mere amusement. The account tells you that after climbing for several hours in the face of novel difficulties, my friend thought it wise to halt so as to secure our retreat; for not one of us knew what difficulties the descent might reveal. I will here state in a few words

Inspection of the Matterhorn 259

what occurred after our separation. Benen and myself had first a hard scramble up some very steep rocks, our motions giving to those below us the impression that we were urging up bales of goods instead of the simple weight of our own bodies. Turning a corner of the ridge we had to cross a very unpleasant-looking slope, the substratum of which was smooth rock, this being covered by about eighteen inches of snow. On ascending, this place was passed in silence, but in coming down the fear arose that the superficial layer might slip away with us; this would hand us over in the twinkling of an eye to the tender mercies of pure gravity for a thousand feet or more. Benen seldom warns me, but he did so here emphatically, declaring his own powerlessness to render any help should the footing give way. Having crossed this slope in our ascent we were fronted by a cliff, against which we rose mainly by aid of the felspar crystals protuberant from its face. Here is the grand difficulty of the Matterhorn; the rocks are sound, smooth, and steep, and hardly offer any grip to either hands or feet. Midway up the cliff referred to, Benen asked me to hold on, as he did not feel sure that it formed the best route. I accordingly ceased moving, and lay against the rock with legs and arms outstretched like a huge and helpless frog. Benen climbed to the top of the cliff, but returned immediately with a flush of confidence in his eye. 'I will lead you to the top,' he said excitedly. Had I been free I should have cried 'Bravo!' but in my position I did not care to risk the muscular motion which a hearty bravo would demand. Aided by the rope I was at his side in a minute, and we soon learned that his confidence was premature. Difficulties thickened round us; on no other mountain are they so thick, and each of them is attended by possibilities of the most blood-chilling kind. Our mode of motion in such circumstances was this:— Benen advanced while I held on to a rock, prepared for the jerk if he should slip. When he had secured himself, he called out, 'Ich bin fest, kommen Sie.' I then worked forward, sometimes halting where he had halted, sometimes passing him until a firm anchorage was gained, when it again became his turn to advance. Thus each of us waited until the other could seize upon something capable of bearing the shock of a sudden descent. At some places Benen deemed a little extra assurance necessary; and here he emphasised his statement that he was 'fest' by a suitable hyperbole. ' Ich bin fest wie ein Mauer,—fest wie ein Berg, ich halte Sie gewiss,' or some such expression. Looking from Breuil, a series of moderate-sized prominences are seen along the *arête* of the Matterhorn; but when you are near them, these black eminences rise like tremendous castles in the air, so wild and high as almost to quell all hope of scaling or getting round them. At the base of one of these edifices Benen paused, and looked closely at the grand mass; he wiped his forehead, and turning to me said, 'Was denken Sie, Herr?'—' Shall we go on, or shall we retreat? I will do what you wish.' 'I am without a wish, Benen,' I replied: 'where you go I follow, be it up or down.' He disliked the idea of giving in, and would willingly have thrown the onus of stopping upon me. We attacked the castle, and by a hard effort reached one of its mid ledges, whence we had plenty of room to examine the remainder. We might certainly have continued the ascent beyond this place, but Benen paused here. To a minute of talk succeeded a minute of silence, during which my guide earnestly scanned the heights. He then turned towards me, and the words seemed to fall from his lips through a resisting medium, as he said, 'Ich denke die Zeit ist zu kurz,'—' It is better to return.' By this time each of the neighbouring peaks had unfolded a cloud banner, remaining clear to windward, but having a streamer hooked on to its summit and drawn far out into space by the moist south wind. It was a grand and affecting sight, grand intrinsically, but doubly impressive to feelings already loosened by the awe inseparable from our position. Looked at from Breuil, the mountain shows two summits separated from each other by a possibly impassable cleft. Only the lower one of these could be seen from our station. I asked Benen how high he supposed it to be above the point where we then stood;

Inspection of the Matterhorn

he estimated its height at 400 feet; I at 500 feet. Probably both of us were under the mark; however, I state the fact as it occurred. The object of my present visit to Breuil was to finish the piece of work thus abruptly broken off, and so I awaited Benen's return with anxious interest.

At dusk I saw him striding down the Alp, and went out to meet him. I sought to gather his opinion from his eye before he spoke, but could make nothing out. It was perfectly firm, but might mean either pro or con. 'Herr,' he said at length, in a tone of unusual emphasis, 'I have examined the mountain carefully, and find it more difficult and dangerous than I had imagined. There is no place upon it where we could well pass the night. We might do so on yonder Col upon the snow, but there we should be almost frozen to death, and totally unfit for the work of the next day. On the rocks there is no ledge or cranny which could give us proper harbourage; and starting from Breuil it is certainly impossible to reach the summit in a single day.' I was entirely taken aback by this report. I felt like a man whose grip had given way, and who was dropping through the air. My thoughts and hopes had laid firm hold upon the Matterhorn, and here my support had suddenly broken off. Benen was evidently dead against any attempt upon the mountain. 'We can, at all events, reach the lower of the two summits,' I remarked. 'Even that is difficult,' he replied; 'but when you have reached it, what then? the peak has neither name nor fame.' I was silent; slightly irascible, perhaps; but it was against the law of my mind to utter a word of remonstrance or persuasion. Benen made his report with his eyes open. He knew me well, and I think mutual trust has rarely been more strongly developed between guide and traveller than between him and me. I knew that I had but to give the word and he would face the mountain with me next day, but it would have been inexcusable in me to deal thus with him. So I stroked my beard, and like Lelia in the 'Princess,' when

'Upon the sward
She tapt her tiny silken-sandal'd foot'

I crushed the grass with my hobnails, seeking thus a safety-valve for my disappointment.

My sleep was unsatisfying that night, and on the following morning I felt a void within. The hope that had filled my mind had been suddenly dislodged, and pure vacuity took its place. It was like the breaking down of a religion, or the removal of a pleasant drug to which one had been long accustomed. I hardly knew what to do with myself. One thing was certain—the Italian valleys had no balm for my state of mind; the mountains alone could restore what I had lost. Over the Joch then once more! We packed up and bade farewell to the host and waiter. Both men seemed smitten with a sudden languor, and could hardly respond to my adieus. They had expected us to be their guests for some time, and were evidently disgusted at our want of pluck. 'Mais, monsieur, il faut faire la pénitence pour une nuit.' I longed for a moment to have the snub-nosed man halfway up the Matterhorn, with no arm but mine to help him down. Veils of the silkiest cloud began to draw themselves round the mountain, and to stretch in long gauzy filaments through the air, where they finally curdled up to common cloud, and lost the grace and beauty of their infancy. Had they condensed to thunder I should have been better satisfied; but it was some consolation to see them thicken so as to hide the mountain, and quench the longing with which I should have viewed its unclouded head. The thought of spending some days chamois hunting occurred to me. Benen seized the idea with delight, promising me an excellent gun. We crossed the summit, descended to Zermatt, paused there to refresh ourselves, and went forward to St. Nicholas, where we spent the night.

CHAPTER XI

OVER THE MORO

' The splendour falls on rocky walls
And snowy summits old in story,
The long light shakes across the lakes,
And the wild cataract leaps in glory.'

BUT time is advancing, and I am growing old; over my left ear, and here and there amid my whiskers, the grey hairs are beginning to peep out. Some few years hence, when the stiffness which belongs to age has unfitted me for anything better, chamois hunting or the Scotch Highlands may suffice; but for the present let me breathe the air of the highest Alps. Thus I pondered on my pallet at St. Nicholas. I had only seen one half of Monte Rosa; and from the Italian side the aspect of the Mountain Queen was unknown to me. I had been upon the Monte Moro three years ago, but looked from it merely into an infinite sea of haze. To complete my knowledge of the mountain it was necessary to go to Macugnaga, and over the Moro I accordingly resolved to go. But resolution had as yet taken no deep root, and on reaching Saas I was beset by the desire to cross the Alphubel. Benen called me at three; but over the pass grey clouds were swung, and as I was determined not to mar this fine excursion by choosing an imperfect day, I then gave it up. At seven o'clock, however, all trace of cloud had disappeared; it had been merely a local gathering of no importance, which the first sunbeams caused to vanish into air. It was now, however, too late to think of the Alphubel, so I reverted to my original design, and at 9 A.M. started up the valley towards Mattmark. A party of friends who were on the road before me contributed strongly to draw me on in this direction.

Onward then we went through the soft green meadows, with the river sounding to our right. The sun showered gold upon the pines, and brought richly out the colouring of the rocks. The blue wood smoke ascended from the hamlets, and the companionable grasshopper sang and chirruped right and left. High up the sides of the mountains the rocks were planed down to tablets by the ancient glaciers. The valley narrows, and we skirt a pile of moraine-like matter, which is roped compactly together by the roots of the pines. Huge blocks here choke the channel of the river, and raise its murmurs to a roar. We emerge from shade into sunshine, and observe the smoke of a distant cataract jetting from the side of the mountain. Crags and boulders are here heaped in confusion upon the hillside, and among them the hardy trees find a lodgment; asking no nutriment from the stones—asking only a pedestal on which they may plant their trunks and lift their branches into the nourishing air. Then comes the cataract itself, plunging in rhythmic gushes down the shining rocks. Rhythm is the rule with Nature;—she abhors uniformity more than she does a vacuum. The passage of a resined bow across a string is typical of her operations. The heart beats by periods, and the messages of sense and motion run along the nerves in oscillations. A liquid cannot flow uniformly through an aperture, but runs by pulses which a little tact may render musical. A flame cannot pass up a funnel without bursting into an organ peal, and when small, as a jet of gas, its periodic flicker can produce a note as pure and sweet as any uttered by the nightingale. The sea waves are rhythmic; and the smaller ripples which form a chasing for the faces of the billows declare the necessity of the liquid to break its motion into periods. Nay, it may be doubted whether the planets themselves move through the space without an intermittent shiver as the ether rubs against their sides. Rhythm is the rule with Nature —

> 'She lays her beams in music,
>   In music every one,
> To the cadence of the whirling world
>   Which dances round the sun.'

Over the Moro

The valley again opens, and finds room for a little hamlet, dingy hovels, with a white little church in the midst of them; patches of green meadow and yellow rye, with the gleam of the river here and there. The moon hangs over the Mischabelhörner, turning a face which ever waxes paler towards the sun. The valley in the distance seems shut in by the Allelein Glacier, towards which we work, amid the waterworn boulders which the river in its hours of fury had here strewn around. The rounded rocks are now beautified with lichens, and scattered trees glimmer among the heaps. Nature heals herself. She feeds the glacier and planes the mountains down. She fuses the glacier and exposes the dead rocks. But instantly her energies are exerted to neutralise the desolation; clothing the crags with splendour, and setting the wind to melody as it wanders through the pines.

At the Mattmark Hotel, which stands, as you perhaps know, at the foot of the Monte Moro, I was joined by a gentleman who had just liberated himself from an unpleasant guide. He was a novice in Switzerland, had been fleeced for a month by his conductor, and finally paid him a considerable sum to be delivered from his presence.[1] Benen halted on the way to adjust his knapsack, while my new companion and myself went on. We lost sight of my guide, lost the track also, and clambered over crag and snow to the summit, where we waited till Benen arrived. The mass of Monte Rosa here grandly revealed itself from top to bottom. Dark cliffs and white snows were finely contrasted, and the longer I looked at it, the more noble and impressive did the mountain appear. We were very soon clear of the snow, and went straight down the declivity towards Macugnaga. There are, or are to be, two hotels at the place, one of which belongs to Lochmatter, the guide. I looked at his house first, but I found a host of men hammering at the stones and rafters. It was still for the most part in a rudimentary

[1] Every class of men has its scoundrels, and the Alpine guides come in for their share. It would be a great boon if some central authority existed, to which cases of real delinquency could be made known.

state. A woman followed us as we receded, and sought to entice Benen back. Had she been clean and fair she might have succeeded, but she was dingy, and therefore failed. We put up at the Monte Moro, where a party of friends greeted me with a vociferous welcome. This was my first visit to Macugnaga, and save as a cauldron for the generation of fogs I knew scarcely anything about it. But there were no fogs there at the time to which I refer, and the place wore quite a charmed aspect. I walked out alone in the evening, up through the meadows towards the base of Monte Rosa, and on no other occasion have I seen peace, beauty, and grandeur so harmoniously blended. Earth and air were exquisite, and I returned to the hotel brimful of delight.

Monte Rosa with her peaks and spurs builds here a noble amphitheatre. From the heart of the mountain creeps the Macugnaga Glacier. To the right a precipitous barrier extends to the Cima di Jazzi, and between the latter and Monte Rosa this barrier is scarred by two couloirs, one of which, or the cliff beside it, has the reputation of forming the old pass of the Weissthor. It had long been a myth whether this so-called 'Alter Pass' had ever been used as such, and many superior mountaineers deemed it from inspection to be impracticable. All doubt on this point was removed this year; for Mr. Tuckett, led by Benen, had crossed the barrier by the couloir most distant from Monte Rosa, and consequently nearest to the Cima di Jazzi. It is a wonder that it had not been scaled by our climbers long ago, for the aspect of the place from Macugnaga is eminently calculated to excite the desire to attack it. As I stood in front of the hotel in the afternoon, I said to Benen that I should like to try the pass on the following day; in ten minutes afterwards, the plan of our expedition was arranged. We were to start before the dawn, and to leave Benen's hands free, a muscular young fellow, who had accompanied Mr. Tuckett, was engaged to carry our provisions. It was also proposed to vary the proceedings by assailing the ridge by the couloir nearest to Monte Rosa.

CHAPTER XII

THE OLD WEISSTHOR

'He lifts me to the golden doors,
The flashes come and go;
All heaven bursts her starry floors
And strows her light below.'

I WAS called by my host at a quarter before three. The firmament of Monte Rosa was almost as black as the rocks beneath it, while above in the darkness trembled the stars. At 4 A.M. we quitted the hotel; a bright half-moon was in the sky, and Orion hung out all his suns. We wound along the meadows, by the slumbering houses, and the unslumbering river. The eastern heaven soon brightened, and we could look direct through the gloom of the valley at the opening of the dawn. We threaded our way amid the boulders which the torrent had scattered over the plain, and among which groups of stately pines now find anchorage. Some of the trees had exerted all their force in a vertical direction, and rose straight, tall, and mast-like without lateral branches. We reached a great moraine, hoary with years, and clothed with magnificent pines; our way lay up it, and from the top we dropped into a little dell of magical beauty. Deep hidden by the glacier-built ridges, guarded by noble trees, soft and green at the bottom, and tufted round with bilberry bushes, through which peeped here and there the lichen-covered crags; I have never seen a spot in which I should so like to dream away a day. Before I entered it, Monte Rosa was still in shadow, but I now noticed that in an incredibly short time all her precipices were in a glow. The purple colouring of the mountains encountered on looking down the valley was indescribable; out of Italy I have never seen anything like it. Oxygen and nitrogen could not produce the effect; some effluence from the earth, some foreign constituent of the atmosphere, developed in those deep valleys by the sun of the south, must sift the solar beams, abstracting a portion, and blending their red and violet to that incomparable hue. In the room where I work in London, there are three classes of actions on calorific rays: the first is due to the pure air itself, the oxygen and nitrogen whose mixture produces our atmosphere; this influence is represented in magnitude by the number 1. A second action is due to the aqueous vapour in the air, and this is represented by the number 40. A third action is due, to what I know not,—but its magnitude is represented by the number 20. As regards, therefore, its action upon radiant heat, the atmosphere of my room embraces a constituent, too minute to be laid hold of by any ordinary method of analysis, and which, nevertheless, is twenty times more potent than the air itself. We know not what we breathe. The air is filled with emanations which vary from day to day, and mainly to such extraneous matters, are the chromatic splendours of our atmosphere to be ascribed. The air south of the Alps is in this respect different from that on the north, but a modicum even of arsenic might be respired with satisfaction, if warmed by the bloom which suffused the air of Italy this glorious dawn.

The ancient moraines of the Macugnaga Glacier rank among the finest that I have ever seen; long, high ridges tapering from base to edge, hoary with age, but beautified by the shrubs and blossoms of to-day. We crossed the ice and them. At the foot of the old Weissthor lay couched a small glacier, which had landed a multitude of boulders on the slope below it; and amid these we were soon threading our way. We crossed the little glacier which at one place strove to be disagreeable, and here I learned from the deportment of his axe the kind of work to which my porter had been previously accustomed. The head of the implement quitted its handle before half-a-dozen strokes had sounded on the ice. We reached the rocks to the right of our couloir and climbed them for some distance. The ice, in fact, at the base of the couloir, was cut by profound fissures, which extended quite across, and rendered a direct advance up the gully impossible. At a proper place we dropped down upon the snow. Close along the rocks it was scarred by a furrow six or eight feet deep, and about twelve in width, evidently the track of avalanches, or of rocks let loose from the heights. Into this we descended. The bottom of the channel was firm and roughened by the stones which found a lodgment there. I thought that we had here a suitable roadway up the couloir, but I had not time to convert the thought into a suggestion, before a crash occurred in the upper regions. I looked aloft, and right over the snow-brow which here closed the view, I perceived a large brown boulder in the air, while a roar of unseen stones showed that the visible projectile was merely the first shot of a general cannonade. They appeared,—pouring straight down upon us,—the sides of the couloir preventing them from squandering their force in any other direction. 'Schnell!' shouted the man behind me, and there is a ring in the word, when sharply uttered in the Alps, that almost lifts a man off his feet. I sprang forward, but urged by a sterner impulse, the man behind sprung right on to me. We cleared the furrow exactly as the first stone flew by, and once in safety we could calmly admire the wild energy with which the rattling boulders sped along.

Our way now lay up the couloir; the snow was steep but knobbly, and hence but few steps were required to give the boots a hold. We crossed and recrossed obliquely, like a laden horse drawing uphill. At times we paused and examined the heights; our couloir ended in the snow-fields above, but near the summit it suddenly rose in a high ice-wall. If we persisted in the couloir, this barrier would have to be surmounted, and the possibility of scaling it was very questionable. Our attention was therefore turned to the rocks at our right, and the thought of assailing them was several times mooted and discussed. They at length seduced us, and we resolved to abandon the couloir. To reach the rocks, however, we had to recross the avalanche channel, which was here very deep. Benen hewed a gap at the top of its flanking wall, and stooping over, scooped steps out of the vertical face of indurated snow. He then made a deep hole in which he anchored his left arm, let himself thus partly down, and with his right pushed the steps to the bottom. While this was going on, small stones were continually flying down the gully. Benen reached the floor and I followed. Our companion was still clinging to the snow wall, when a horrible clatter was heard overhead. It was another stone avalanche, which there was hardly a hope of escaping. Happily a rock was here firmly stuck in the bed of the gully, and I chanced to be beside it when the first huge missile appeared. This was the delinquent which had set the others loose. I was directly in the line of fire, but ducking behind the boulder I let the projectile shoot over my head. Behind it came a shoal of smaller fry, each of them, however, quite competent to crack a human life. Benen shouted 'Quick!' and never before had I seen his axe so promptly wielded. You must remember that while this infernal cannonade was being executed, we hung upon a slope of snow which had been pressed and polished to ice by the descending stones; and so steep that a single slip would have converted us into an avalanche also. Without steps of some kind we dared not set foot on the slope, and these had to be cut while the stone shower was in the act of falling on us. Mere scratches in the ice, however, were all the axe could accomplish, and on these we steadied ourselves with the energy of desperate men. Benen was first, and I followed him, while the stones flew thick beside and between us. Once an ugly lump made right at me; I might perhaps have dodged it, but Benen saw it coming, turned, caught it on the handle of his axe as a cricketer catches a ball, and thus deflected it from me. The labour of his axe was here for a time divided between the projectiles and the ice, while at every pause in the volley, 'he cut a step and sprang forward.' Had the peril been less, it would have been amusing to see our contortions as we fenced with our swarming foes. A final jump landed us on an embankment, out of the direct line of fire which raked the gully, and we thus escaped a danger new in this form and extremely exciting to us all. We had next to descend an ice slope to the place at which the rocks were to be invaded. Andermatten slipped here, shot down the slope, knocked Benen off his legs, but before the rope had jerked me off mine, Benen had stopped his flight. The porter's hat, however, was shaken from his head and lost. Our work, as you will see, was not without peril, but if real discipline for eye, limb, head, and heart, be of any value, we had it here.

Behold us then fairly committed to the rocks; our first acquaintance with them was by no means comforting,—they were uniformly steep, and as far as we could judge from a long look upwards they were likely to continue so. A stiffer bit than ordinary interposed now and then, making us feel how possible it was to be entirely cut off. We at length reached real difficulty number one: all three of us were huddled together on a narrow ledge, with a smooth and vertical cliff above us. Benen tried it in various ways while we held on to the rocks, but he was several times forced back to the ledge. At length he managed to get the fingers of one hand over the top of the cliff, while to aid his grip he tried to fasten his shoes against its face. But the nails scraped freely over the granular surface, and he had practically to lift himself by a single arm. As he did so he had the ugliest place beneath him over which a human body could well be suspended. We were tied to him of course; but the jerk, had his grip failed, would have been terrible. I am not given to heart-beat, but here my organ throbbed a little. By a great effort he raised his breast to a level with the top, and leaning over it he relieved the strain upon his arm. Supported thus he seized upon something further on, and lifted himself quite to the top. He then tightened the rope, and I slowly worked myself over the face of the cliff after him. We were soon side by side, while immediately afterwards Andermatten, with his long unkempt hair, and face white with excitement, hung midway between heaven and earth supported by the rope alone. We hauled him up bodily, and as he stood upon our ledge, his limbs quivered beneath him.

We now strained slowly upwards amid the maze of crags, and scaled a second cliff resembling, though in a modified form, that just described. There was no peace, no rest, no delivery from the anxiety 'which weighed upon the heart.' Benen looked extremely blank, and often cast an eye downward to the couloir, which we had quitted, muttering aloud, 'Had we only stuck to the snow!' He had soon reason to emphasise his ejaculation. After climbing for some time, we reached a smooth vertical face of rock from which right or left, there was no escape, and over which we must go. Benen first tried it unaided, but was obliged to recoil. Without a lift of five or six feet, the thing was impossible. When a boy I have often climbed a wall by placing a comrade in a stooping posture with his hands and head against the wall, getting on his back, and permitting him gradually to straighten himself till he became erect. This plan I now proposed to Benen, offering to take him on my back. 'Nein, Herr!' he replied; 'nicht Sie, ich will es mit Andermatten versuchen.' I could not persuade him, so Andermatten got upon the ledge, and fixed his knee for Benen to stand on. In this position my guide obtained a precarious grip, just sufficient to enable him to pass with safety from the knee to the shoulder. He paused here, and pulled away such splinters as might prove treacherous, if he laid hold of them. He at length found a firm one, and had next to urge himself, not fairly upward, for right above us the top was entirely out of reach, but obliquely along the face of the cliff. He succeeded, anchored himself, and called upon me to advance. The rope was tight, it is true, but it was not vertical, so that a slip would cause me to swing like a pendulum over the cliff's face. With considerable effort I managed to hand Benen his axe, and while doing so my own staff escaped me and was irrecoverably lost. I ascended Andermatten's shoulders as Benen did, but my body was not long enough to bridge the way to Benen's arm; I had to risk the possibility of becoming a pendulum. A little protrusion gave my left foot some support. I raised myself a yard, and here was suddenly met by the iron grip of my guide. In a second I was safely stowed away in a neighbouring fissure. Andermatten now remained. He first detached himself from the rope, tied it round his coat and knapsack which were drawn up. The rope was again let down, and the porter tied it firmly round his waist, it tightened and lifted him tiptoe. It was not made in England, and was perhaps lighter than it ought to be; to help it hands and feet were scraped with spasmodic energy over the rock. He struggled too much, and Benen cried sharply, and apparently with some anxiety, 'Langsam! langsam! Keine Furcht!' The poor fellow looked very pale and bewildered as his bare head emerged above the ledge. His body soon followed. Benen always uses the imperfect for the present tense, 'Er war ganz bleich,' he remarks to me, the 'war' meaning *ist*.

The young man seemed to regard Benen with a kind of awe. 'Mein Herr,' he exclaimed, 'you would not find another guide in Switzerland to lead you up here.' Nor, indeed, to Benen's credit be it spoken, would he have done so if he could have avoided it; but we had fairly got into a net, the meshes of which must be resolutely cut. I had previously entertained the undoubting belief that where a chamois could climb a man could follow; but when I saw the marks of the animal on these all but inaccessible ledges, my belief, though not eradicated, became weaker than it had previously been. Onward again slowly winding through the craggy mazes, and closely scanning the cliffs as we ascended. Our easiest work was stiff, but the 'stiff' was an agreeable relaxation from the perilous. By a lateral deviation we reached a point whence we could look into the couloir by which Mr. Tuckett had ascended: here Benen relieved himself by a sigh and ejaculation: 'Would that we had chosen it, we might pass up yonder rocks blindfold!' But repining was useless, our work was marked out for us and must be accomplished. After another difficult tug Benen reached a point whence he could see a large extent of the rocks above us. There was no serious difficulty within view, and the announcement of this cheered us mightily. Every vertical yard, however, was to be won only by strenuous effort. For a long time the snow cornice hung high above us; we now approach its level; the last cliff forms a sloping stair with strata for steps. We spring up it, and the magnificent snow-field of the Görner Glacier immediately opens to our view. The anxiety of the last four hours disappears like an unpleasant dream, and with that perfect happiness which perfect health can alone impart, we consumed our cold mutton and champagne on the summit of the old Weissthor.

To the habits of the mountaineer Milton's opinion regarding the utility of teaching the use of weapons to his pupils is especially applicable. Such exercises constitute 'a good means of making them healthy, nimble, and well in breath, and of inspiring them with a gallant and fearless courage, which, being tempered with seasonable precepts of true fortitude and patience, shall turn into a native and heroic valour, and make them hate the cowardice of doing wrong.' Farewell!

THE END

COSIMO CLASSICS

COSIMO is an innovative publisher of books and publications that inspire, inform and engage readers worldwide. Our titles are drawn from a range of subjects including health, business, philosophy, history, science and sacred texts. We specialize in using print-on-demand technology (POD), making it possible to publish books for both general and specialized audiences and to keep books in print indefinitely. With POD technology new titles can reach their audiences faster and more efficiently than with traditional publishing.

> **Permanent Availability:** Our books & publications never go out-of-print.

> **Global Availability:** Our books are always available online at popular retailers and can be ordered from your favorite local bookstore.

COSIMO CLASSICS brings to life unique, rare, out-of-print classics representing subjects as diverse as *Alternative Health, Business and Economics, Eastern Philosophy, Personal Growth, Mythology, Philosophy, Sacred Texts, Science, Spirituality* and much more!

COSIMO-on-DEMAND publishes your books, publications and reports. If you are an Author, part of an Organization, or a Benefactor with a publishing project and would like to bring books back into print, publish new books fast and effectively, would like your publications, books, training guides, and conference reports to be made available to your members and wider audiences around the world, we can assist you with your publishing needs.

Visit our website at www.cosimobooks.com to learn more about Cosimo, browse our catalog, take part in surveys or campaigns, and sign-up for our newsletter.

And if you wish please drop us a line at info@cosimobooks.com. We look forward to hearing from you.

www.ingramcontent.com/pod-product-compliance
Lightning Source LLC
Chambersburg PA
CBHW032102090426
42743CB00007B/208